Separating Pro-Environment Technologies for Waste Treatment, Soil and Sediments Remediation

Editor

Vincenzo Gente

Environmental Engineer
Italy

Co-Editor

Floriana La Marca

University of Rome "La Sapienza"
Rome
Italy

CONTENTS

Foreword   *i*

Preface   *iii*

List of Contributors   *v*

CHATPERS

1. **Waste Treatment Based on Mineral Processing Techniques**   3
   M. Ghiani and A. Zucca

2. **Challenges in the Separation of Plastics from Packaging Waste**   30
   M.T. Carvalho

3. **ADR: A Classifier for Fine Moist Materials**   43
   W. de Vries and P.C. Rem

4. **Sensor Based Sorting in Waste Processing**   59
   J. Julius and Th. Pretz

5. **Upgrading of Post-Consumer Steel Scrap**   77
   F. Quarta, A. Bonoli and P.C. Rem

6. **State of the Art of Recycling of Photovoltaic Panels Using Separation Technology**   90
   A. Bonoli and A. Pompei

7. **The Control of Separation Processes in Mechanical Recycling of Waste Refrigerators by Partition Function**   109
   F. La Marca

8. **A Matrix Based Approach for Modelling of Treatment Processes for Contaminated Groundwater**   123
   V. Gente and D. Lausdei

9. **Separation Technologies for Inorganic Compounds Contained in Industrial Wastewaters Including Metal Ions, Metalloids, Thiosalts, Cyanide, Ammonia and Nitrate**   **139**

     N. Kuyucak and I.Toreci Mubarek

10. **Evaluation of Sediments' Contamination by Hyperspectral Analysis**   **172**

     V. Gente, S. Geraldini, F. La Marca and F. Palombo

Index   **188**

FOREWORD

The eBook entitled «Separating Pro-Environment-Technologies for waste treatment and soil and sediments remediation» aims to underline, through several well selected and presented case studies from waste and groundwater treatment as well as soil and sediment remediation, that well established mineral processing technologies may be used with appropriate modifications for hazardous waste management and the recovery of metals, including rare and precious, and metalloids.

M. Ghiani and A. Zucca explore the potential of established mineral processing techniques, such as size classification, attrition, gravity separation and flotation for contaminated waste and dredged sediment management and the recovery of marketable products.

M.T. Carvalho discusses the application of gravity concentration and froth flotation to meet the challenges in separation of plastics from packaging material and highlights advantages and limitations.

W. de Vries and P.C. Rem assess the performance of a new type of classifier, called Advanced Dry Recovery (ADR), which is able to separate moist fines from Incinerator Bottom Ash (IBA) and Construction and Demolition Waste (CDW).

J. Julius and Th. Pretz focus on the advantages of sensor based sorting to recognize multiple material properties and its potential applications in waste processing.

A. Quarta, A. Bonoli and P.C. Rem present the advantages of a shape-sensitive magnetic separator called "Clean Scrap Machine" (CSM) that pre-sorts scrap into a bulky fine steel chips fraction of high purity.

A. Bonoli and A. Pompei present the state of the art of recycling photovoltaic panels using separation technologies which enable recovery of metals as well as other materials.

F. La Marca proposes a mathematical model that can be implemented during recycling of waste refrigerators to define a partition function, utilizing data obtained from the treatment of waste refrigerator carcasses. The partition function determines control parameters and is therefore able to assess the quality of the recovered products, and, the efficiency of the applied technologies.

V. Gente and D. Lausdei present a matrix approach for modelling of a pump and treat system that is used for groundwater decontamination. The adopted approach determines the recirculation ratio of groundwater so that its final quality can meet discharge criteria.

N. Kuyucak and I. Toreci Mubarek discuss conventional and emerging separation technologies that can be used for the removal or recovery of metals, metalloids, thiosalts, cyanide, ammonia and nitrate from industrial wastewaters.

Finally, V. Gente, S. Geraldini, F. La Marca and F. Palombo propose a new approach based on hyperspectral analysis for a reliable evaluation of contaminated dredged sediments that determines their appropriate management according to standards and regulations.

All case studies included in this eBook are of great interest to engineers, scientists from national and regional authorities as well as to SMEs/industry, researchers and consultants. This eBook is also useful for post graduate students who are involved in environmental projects. The proposed technologies contribute to waste valorisation, prevention of depletion of natural resources and minimization of environmental impacts. These technologies should also be considered during LCA studies by adopting well established technical, economic, environmental and social criteria.

The efforts of the editor Vincenzo Gente and the co-editor Floriana La Marca for the selection of the case studies and the production of this eBook are highly acknowledged.

Prof. Konstantinos Komnitsas

Technical University Crete
Dpt. Mineral Resources Engineering
Chania
Greece
Email: komni@mred.tuc.gr

PREFACE

When dealing with environmental issues, most of the times it is necessary to deal with contaminants or pollutants, which have to be removed from waste materials or from natural matrices.

This is the case, for example, during recovery of high-added value constituents from waste streams for recycling or reuse, and extraction (or removal) of contaminants from soil or sediments.

In both cases, the aim of the actions that have to be taken in order to tackle the environmental issues, is to separate what is useful (recyclable materials, soil and sediments) from what is useless (non recyclable materials, contaminants).

A close look at the technologies developed in mineral processing could be of great help in finding the right solution. As a matter of fact, the liberation and separation processes used for the recovery of valuable minerals from gangue are principally the same processes that can be applied to waste in order to recover valuable materials as well as to soil and sediments to eliminate contamination.

During the last decades, the raising concern in environmental issues has led to the adoption of technologies, initially developed for mineral processing. Processes that include comminution, classification or electrical and magnetic separation, hydrocycloning and flotation, can also be considered for the treatment of either waste materials or remediation of soils and sediments.

Nevertheless, prior to the use of mineral technologies in environmental applications some modifications and tests are required in order to optimize treatment efficiency by considering technical, economic, environmental and social criteria.

In fact, the differences between ore deposits and waste materials, soils and sediments as well as their specific physico-chemical characteristics in each case, have to be taken into consideration:

- Ore deposits are usually more homogeneous compared to waste materials and contaminated soils and sediments, that are often heterogeneous in terms of size distribution, and composition;

- The composition of ore deposits is usually more stable while the composition of wastes, soils and sediments may vary with time even in nearby sites; in the latter case hotspots are often seen, thus making their treatment more complex;

- The characteristics of concentrates produced during mineral processing are well specified by market standards, whereas valuable constituents that can be recovered from waste are not always specific; furthermore the establishment of different thresholds in each Member State and the plethora of legislation, beyond European Commission Directives and Best Available Techniques, for specific contaminants may impose the adoption of different treatment options.

Therefore, the aim of this eBook is to investigate how mineral processing technologies, can be modified and improved when applied to waste treatment as well as to soil and sediment remediation.

In particular, emphasis is placed on the application of different technologies to different matrices as well as the innovations foreseen from their use in environmental applications and in specific industrial sectors.

Vincenzo Gente
Environmental Engineer
Italy

&

Floriana La Marca
University of Rome "La Sapienza"
Rome
Italy

List of Contributors

Bonoli A.

Department of Civil Environment Materials Engineering, University of Bologna, Italy
E-mail: alessandra.bonoli@unibo.it

Carvalho M.T.

CERENA, Instituto Superior Técnico, Lisboa, Portugal
E-mail: teresa.carvalho@ist.utl.pt

de Vries W.

Department of Resoucrces & Recycling, Delft University of Technology, The Netherlands
E-mail: W.deVries@tudelft.nl

Gente V.

Environmental Engineer, Italy
E-mail: vincenzo.gente@ingpec.eu

Geraldini S.

Istituto Superiore per la Protezione e Ricerca Ambientale, Rome, Italy
E-mail: serena.geraldini@isprambiente.it

Ghiani M.

Department of Geoengineering and Environmental Technologies, University of Cagliari, Italy
E-mail: ghiani@unica.it

Julius J.

Department of Processing and Recycling RWTH Aachen University, Aachen, Germany
E-mail: julius@ifa.rwth-aachen.de

Kuyucak N.

Golder Associates Paste Technology Ltd., Ontario, Canada
E-mail: NKuyucak@golder.com

La Marca F.

Department of Chemical Engineering Materials & Environment, University of Rome "La Sapienza", Rome, Italy
E-mail: Floriana.lamarca@uniroma1.it

Lausdei D.

ENVIRON Italy S.r.l., Rome, Italy
E-mail: dlausdei@environcorp.com

Palombo F.

Golder Associates S.r.l., Rome, Italy
E-mail: fpalombo@golder.com

Pompei A.

Department of Civil Environment Materials Engineering, University of Bologna, Italy

Pretz Th.

Department of Processing and Recycling RWTH Aachen University, Aachen, Germany
E-mail: pretz@ifa.rwth-aachen.de

Quarta F.

Department of Civil Environment Materials Engineering, University of Bologna, Italy

Rem P.C.

Department of Resoucrces & Recycling, Delft University of Technology, The Netherlands
E-mail: P.C.Rem@tudelft.nl

Toreci Mubarek I.

Golder Associates Ltd., Ontario, Canada
E-mail: IMubarek@golder.com

Zucca A.

Environmental Geology and Geoengineering Institute of CNR, UOS of Cagliari, Italy

E-mail: azucca@unica.it

CHAPTER 1

Waste Treatment Based on Mineral Processing Techniques

M. Ghiani[1] and A. Zucca[2,*]

[1]*Department of Geoengineering and Environmental technologies, University of Cagliari, Italy and* [2]*Environmental Geology and Geoengineering Institute of CNR, UOS of Cagliari, Italy*

Abstract: The use of mineral processing techniques, such as size classification, attrition, gravity separation, flotation *etc.*, can be considered attractive to treat contaminated waste when there is the possibility to recover marketable materials and/or to reduce drastically the volume and the contaminants of the rejects to be located in controlled waste disposal sites. This paper gives some examples of their application to treat: a mine waste contaminated with heavy metals; a granite waste, a contaminated sediments dredging and fly ash from a coal power station.

Keywords: Mineral processing, Heavy metal mine waste, Granite waste, Dredging waste, Fly ash.

INTRODUCTION

The decline of traditional industries and especially of mining industry in industrialized countries have to be compensated by alternative economic activities in the affected areas, with the result that existing or potential sources of pollution have to be eliminated or rendered non-hazardous. Remediating land contaminated by past industrial activities is often a complex and costly process. This is the main reason why the opening of new mines is nowadays subject to strict regulations. Accordingly, in many countries the preliminary administrative authorisation requires a careful environmental impact assessment; therefore in the design of new mine activities it is necessary to adopt suitable solutions aimed at minimising the production and the propagation of contaminants in the environment.

Waste treatment, aimed at reducing the presence of contaminants in industrial rejects and soil remediation, involves a number of chemical, physical and physico-

---
*Address correspondence to A. Zucca: Environmental Geology and Geoengineering Institute of CNR, UOS of Cagliari, Piazza d 'Armi 19, Cagliari, Italy; Tel: +390706755504; Fax: +39 070.6755523; Email: azucca@unica.it

chemical treatment processes based on traditional mineral processing techniques. The authors will consider in particular solid/liquid and solid/solid separation techniques, both used to eliminate or to drastically reduce the content of heavy metals in waste to be landfilled and to recover useful and marketable materials.

If the pollution due to the presence of heavy metals, rather usual in industrial areas (mine areas in particular), is taken into consideration, it can be noted that it occurs in different forms such as: metal ions adsorbed in more or less porous matrixes; original or partially altered mineral phase (*e.g.* blende, galena, chalcopyrite *etc.*); and products or secondary minerals (*e.g.* metal carbonates, sulphates and hydroxides) due to dissolution/precipitation phenomena that constitute a distinct mineralogical phase coating soil particles. The characteristics of the contaminated waste, as well as the grade and the characteristic of pollution, will drive the choice of the suitable treatment or combination of treatments to be adopted.

When the pollution is due to adsorption phenomena, a vigorous washing of the soil - using suitable machines such as attrition cells or log washers - with water solutions and the addition of extracting agents, in order to help *the transfer of the contaminants to the liquid phase,* can be considered a suitable solution. When the transfer of the contaminants to the liquid phase is effective, the following step consisting of the separation of the liquid from the solid phase, using *dewatering techniques such as* filtration and centrifugation or cyclone techniques, can permit achieving the objective of the soil remediation. A further contribution to minimize the content of contaminants in the solid phase can be pursued by the elimination of the clay fractions using classification techniques.

When the pollution is due to the presence of original mineral phases, solid/solid separation techniques can be potentially applied. When dealing with mineral phases containing heavy metals, different techniques can be applied, separately or in combination, depending on the characteristics of the "raw material", such as: size classification techniques, when the contaminants are confined in defined size classes, finest size fractions for instance; gravity techniques, when separation can be achieved through the difference in specific gravity of the mineral species; magnetic and electric separation techniques, for separating particles of different magnetic or electrostatic characteristics, and; froth flotation techniques for separating hydrophobic from hydrophilic particles.

When the pollution is due to the products of dissolution/precipitation phenomena, generally coating soil particles, the soil remediation can be achieved by combining different technologies. A vigorous attrition followed by cyclone dewatering and size classification together with solid/solid techniques applied on the underflow of the cyclone could be a suitable sequence to achieve the objective.

When the soils are polluted both by acid drainage and by the migration of solid contaminant from dumps and slag heaps, due to lack of maintenance and safety measures, the contamination – with the contemporary presence of adsorption phenomena, original mineral phases and products of dissolution/precipitation – may be rather complex and onerous to solve using the above mentioned separation methods.

A careful comparison of the results obtained with traditional mineral processing techniques and with *in situ* techniques – such as adding suitable additives to immobilise heavy metals, use of Permeable Reactive Barriers (PRBs) and filters, use of phytoremediation and electrokinetics – together with the cost-benefit evaluation, can drive the final choice. The use of mineral processing techniques can be considered particularly attractive for treatment of sources of hazardous pollutants like old mine dumps and spoil heaps, where "poor materials" – meaning those coming from mine preparation works carried out on the part of the deposit with a low content of useful minerals and considered, during the period of exploitation, not suitable to obtain marketable products – or rejects from low efficiency treatment have been disposed off. In these cases, the new developments in the solid/solid separation techniques can obtain satisfactory results both to minimize the environmental pollution and to recover useful materials. It is necessary to point out that the application of mineral processing techniques in site remediation – contrary to what happens in the traditional mineral processing plants – must aim at obtaining a product with the minimum content of contaminants, according to legislative limits and generally-accepted practices, and at minimizing the volume of contaminated material to be confined in suitable sites. Additionally, the possibility to recover useful products must be taken into account to obtain marketable concentrates from waste which is generally desirable and can be helpful in reducing the onerous process costs.

Concerning treatment techniques based on traditional mineral processing, the paper reports the following studies of DIGITA (Department of Geoengineering and Environmental technologies University of Cagliari, Italy) together with IGAG (Environmental Geology and Geoengineering Institute of CNR, UOS of Cagliari, Italy).

Mine Waste Contaminated with Heavy Metals (The Case of Monte Narba Mine Site)

The Monte Narba site (south-east Sardinia, Italy) is situated in a landscape of great natural beauty with its steep rugged slopes, narrow valleys, rocky crests jutting out from craggy rock walls, covered with locally a dense vegetation of shrubs and trees.

The lead and silver-rich deposits of Monte Narba were exploited for some thirty years in the late 19th century but productivity rapidly declined in the early 1900's until the mine was eventually shut down in 1933. There are 9 mine spoil heaps situated along the hillsides surrounding the site and another 3 waste dumps at the foot of the main watersheds. The mine waste covers a total area of some 30,000 m^2 amounting to a volume of 298,000 m^3, 215,000 of which are contained in the hillside heaps and 83,000 m^3 are deposited along the main valley-side gullies. The lack of post-closure maintenance has aggravated the problem of erosion and during the severe rain storms that occur periodically, huge quantities of mine waste are transported from the spoil heaps and deposited down-gradient in the beds of two streams. This material has completely silted up a dam erected up-gradient from the site and partially buried most of the mine buildings in the vicinity of the stream beds.

The geochemical investigation revealed elevated metal content in the spoil dumps as well as in the stream sediments, ranging from 800-11,600 ppm Pb, and 1,500-4,000 ppm Zn (Progemisa, 2000). Leaching tests conducted on the above materials in accordance with Italian standards indicated heavy metal release exceeding regulatory limits (D.Lgs. n. 152, 3 April 2006). The considerable amounts of mine waste prone to erosion and transport constitute a potential pollution hazard in the down-gradient flood plain where crops and cereals are cultivated. Moreover, huge quantities of the mine waste, that were transported down-gradient by flood waters, must be removed before restoration of the mine buildings and urban remediation of the site for possible use as a museum or as a tourist attraction can be considered.

Two possible solutions to the problem have been analysed:

- Removal of the waste material and haulage to controlled waste disposal sites;

- Remediation of the removed material to obtain a low–polluting "product" to be relocated or used as aggregate.

This latter solution, applying mineral processing techniques, has been investigated (De Carlo *et al.*, 2001).

A sample composed of the mine waste transported by flood waters was collected by means of chip channels and from small trenches dug along the beds of two streams down-gradient from the site and from the dam up-gradient of the mining village.

Macroscopic observation of the samples and examination under the binocular microscope indicated the following lithology and mineralogical composition:

- Host rock: black schist, metasandstone, quarzite, grey porphyry, granitoid rock;

- Mineralization: large crystals of galena, reddish brown sphalerite, pyrite in the quartz and schist, pyrrhotite, compact and banded quartz, compact white barytine, compact and rare crystalline calcite, rare light green fluorite.

The sample was crushed to below 10 mm and divided into the grain size fractions used in the experiments for determining washability and in the ore dressing tests.

Table **1** shows the results of size analysis, after reduction and classification, and of chemical analysis for each size class.

Table **2** gives the specific gravity analysis in heavy liquids for the –10+5 mm and –5+1 mm size classes respectively.

Table 1: Size and chemical analysis of the sample after reduction to -10 mm

Class	Weight	Assay (ppm)						
(mm)	(%)	Ag	As	Ba	Cd	Cu	Pb	Zn
-10+5	38.3	23.5	92.9	5693.8	14.8	91.2	11,304.4	3,186.8
-5+1	37.9	42.2	82.8	10,955.7	10.8	84.1	8,830.0	2,848.7
-1+0.5	10.2	131.5	86.2	13,570.7	12.1	84.2	9,899.1	3,496.5
-0.5	13.6	87.1	131.2	16,820.1	21.4	128.8	9,895.3	5,801.3
Total	100.0	50.3	93.6	10,004.7	13.9	92.9	10,031.6	3,445.8

Table 2: Gravity analysis on the –10+5 mm e –5+1 mm size classes

Gravity	Weight	Assay (ppm)						
(g/cm^3)	(%)	Ag	As	Ba	Cd	Cu	Pb	Zn
–10+5 mm Size Class								
-2.60	12.4	3	77	1,688	8	69	658	1,745
+2.60-2.65	29.9	3	76	809	8	70	599	1,598
+2.65-2.70	33.6	3	66	724	12	86	1,147	2,691
+2.70-2.80	16.1	4	45	770	11	104	1,128	2,413
+2.80-2.89	3.8	8	93	1,179	44	253	3,881	10,190
+2.89	4.2	543	372	112,773	109	339	237,800	18,520
Total	100.0	25.8	80.7	5,551.9	15.4	98.9	10,861.3	3,145.4
–5+1 mm Size Class								
-2.60	13.4	5	38	1,586	9	71	635	2,670
+2.60-2.65	36.6	3	43	682	6	52	673	1,915
+2.65-2.70	25.7	6	40	592	8	63	933	2,223
+2.70-2.80	15.3	10	45	627	12	101	1,467	2,965
+2.80-2.89	3.2	45	93	1,152	26	161	2,625	5,483
+2.89	5.7	646	527	127,110	58	396	142,900	11,530
Total	100.0	43.2	71.1	8,001.1	11.4	88.0	9,035.3	2,920.2

The experimental results demonstrate the amenability of the mine waste to gravity separation. Assuming a specific gravity of separation of 2.80 g/cm^3, a pre-concentrate can theoretically be obtained with Ag, Ba and Pb recoveries of around 90%.

The Pb and Ag concentrations are comparable with the run-of-mine currently supplied to ore processing plants and the heavy metal concentrations of the reject fraction comply with Italian regulatory limits for sites for commercial or industrial use, with the exception of zinc (regulatory limit value: 1500 ppm).

Tests with Jigs and Shaking Tables

Laboratory tests were conducted using jigs for the -10+1 mm size class and shaking tables for the -1 mm size class.

Table **3** shows the results for the jigs and shaking tables as well as the overall results for a flow sheet incorporating the two processes.

Table 3: Results of jig and shaking table tests

Products	Weight (%)	Assay (ppm)						
		Ag	As	Ba	Cd	Cu	Pb	Zn
Jigs, -10+1 mm Size Class								
Preconcentrate	12.8	156.2	280.8	51,057	24.9	144.3	57,231	5,594
Reject	63.4	6.1	52.5	1,555	9.7	86.4	965	2,482
Total	76.2	31.3	90.8	9,870	12.2	96.1	10,417	3,006
Shaking Tables, -1 mm Grain Class								
Preconcentrate	2.9	775.1	411.7	114,549	60.8	264.0	73,329	11,169
Reject	20.9	17.7	66.2	1,857	11.7	89.2	1138	3,900
Total	23.8	110.0	108.3	15,589	17.7	110.5	9934	4,786
Overall Results								
Preconcentrate	15.7	270.2	304.5	62,696	31.4	166.1	60,105	6,615
Reject	84.3	8.9	55.9	1,630	10.2	87.1	1,008	2,835
Total	100.0	50.0	95.0	11,231	13.5	99.5	10,302	3,429

The results obtained confirm that the mine waste can be satisfactorily processed *via* gravity separation obtaining two products at relatively low costs:

- A pre-concentrate containing 6.02% Pb, 270 g/t Ag and 6.3% $BaSO_4$, suitable for further processing in an ore dressing plant;

-      A reject fraction with concentrations of potentially harmful minerals that comply with current regulatory limits for "sites for commercial and industrial use" (D.Lgs. n. 152, 3 April 2006), with the exception of zinc.

Applying only gravity separation techniques, it is not possible to reduce the Zn content to below the regulatory level of 1500 ppm for "commercial and industrial land use", but leaching tests conducted in accordance with Italian standards, on the total reject indicated heavy metal release does not exceed regulatory limits.

To complete the investigation, both pre-concentrate and reject of gravity separation tests have been further processed following the flow sheet shown in Fig. **1**.

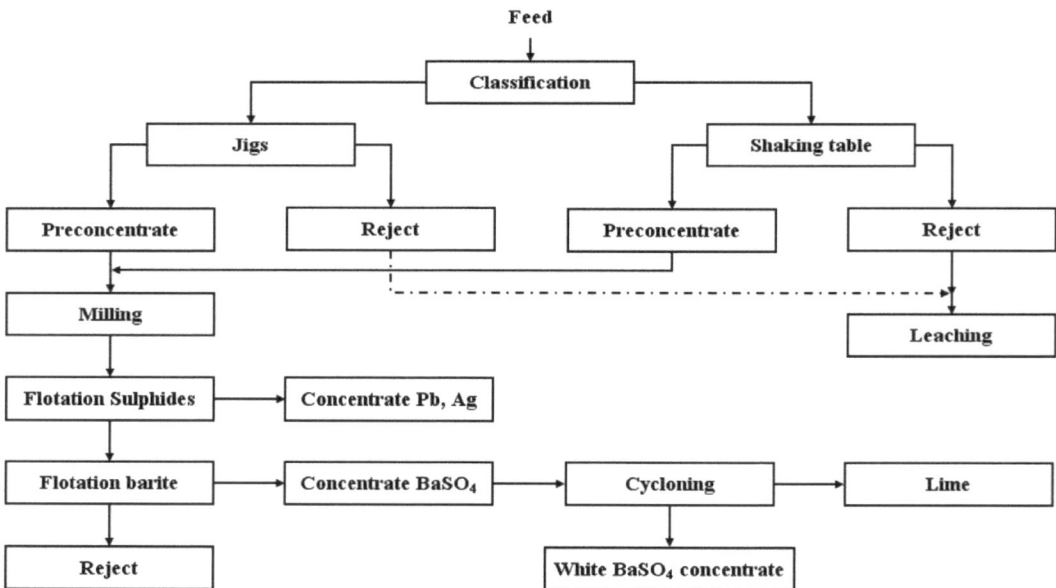

Figure 1: Flow sheet.

To evaluate the possibility of a further reduction of contaminants, the reject obtained from shaking table has been leached using a EDTA 0.2 M solution and HCl (0.2 M) + $CaCl_2$ (4 M Cl⁻) brine. The results of leaching tests (Table **4**) show that, using both EDTA solution and HCl (0.2 M) + $CaCl_2$ brine (this later with more effectiveness), it is possible to reduce contaminants below the limits of Italian regulation for sites of commercial or industrial use (D.Lgs. 152/06, Table 1B).

Table 4: Results of leaching tests carried out on the shaking table reject

Heavy Metals	Pb (ppm)	Zn (ppm)	Cu (ppm)	Cd (ppm)
EDTA 0.2 M	118	1,464	52	2
HCl-CaCl2	99	260	47	0.4
D.Lgs. 152/06, Table 1B	1,000	1,500	600	15

To obtain a suitable degree of liberation, the pre-concentrate product of gravity separation was milled below 0.15 mm before sulphide flotation.

Sulphide flotation was performed with one roughing and two cleaning stages using potassium amyl-xanthate (KAX) as collector agent and Methyl Isobutyl Carbinol (MIBC) as frothing agent.

A concentrate assaying more than 73% Pb and 4500 ppm Ag was obtained and the recovery of the two elements was respectively 95.7 and 88.0% referring to the pre-concentrate. $BaSO_4$ flotation, fed with the reject of sulphide flotation (assaying 10.7% $BaSO_4$) and performed with one roughing and four cleaning stages, permits to obtain a concentrate assaying more than 98% $BaSO_4$ with a recovery of 88% referring to the feed of barite flotation. Na cetylsulphate as barite collector, Na-silicate as peptising agent and NaOH as pH modifier have been used.

To improve the whiteness of $BaSO_4$ concentrate, a cycloning stage permits to obtain a marketable (-20 μm) overflow and an underflow BaSO4 product with 91% whiteness degree.

Based on the indications emerging from the cleaning tests and bearing in mind that the ore dressing techniques tested are designed to clean up the site "applying the best technologies available at reasonable costs", the results obtained demonstrate that the mine waste can be satisfactorily processed *via* gravity separation obtaining two products at relatively low costs:

- A pre-concentrate suitable for further processing in ore dressing plants, thereby minimizing treatment costs and reducing the amount of material to be incorporated into the existing spoil heaps;

- A reject fraction with harmful mineral concentrations falling within the current regulatory limits, with the exception of zinc, for "sites for commercial and industrial use " (D.Lgs. n. 152, 3 April 2006), which could be, in this case, disposed of within the mine site itself.

The re-profiled dumps could then be re-vegetated for the dual purpose of protecting against erosion by storm runoff and further eliminating residual toxic substances while employing phyto-remediation techniques.

MINE WASTE CONTAMINATED WITH HEAVY METALS (THE CASE OF MONTEVECCHIO MINE SITE)

Montevecchio-Levante mine (located in southern-western Sardinia) has been closed in 1994. Mining activities thrived for one hundred and fifty years during which about sixty million tons have been exploited with Pb+Zn average content of 11%. The presence of spoil heaps and of a large mine tailings pond has led to severe degradation of the floodplain. The contamination is the result, on the one hand, of the practice in the past (up to the 1950's) of systematically opening tailings ponds embankments and, on the other hand, of exceptional flood events (in November 1933 for example, mean precipitation amounted to some 484 mm). Protests by farmers against the mining company are documented since the 1930s.

Investigations conducted so far for the purpose of assessing the magnitude and extent of soil contamination have indicated extremely high concentrations of metals such as Pb (120-11,350 ppm), Zn (140-11,400 ppm), Cd (3-98 ppm) and As (18-1,180 ppm) and a volume in the order of $2 \cdot 10^6$ m^3 of more or less contaminated sediments composed of fines transported from the washing plant. These are distributed along the entire 25 km course of the Montevecchio-Sitzerri river that flows downstream from the Montevecchio-Levante mine tailings settling pond.

The thickness of the sediments and soils ranges from 20 cm up to 2 m and over the years they have been variously reworked and transported also by farmers ploughing further reject of the size fraction up to 0.5 mm always by means of screen classification in an attempt to recover agricultural land (PROGEMISA SpA, 1998).

The possibilities of soil decontamination using washing and leaching techniques have been investigated (Dessì *et al.*, 1999; Dessì *et al.*, 2000; Dessì *et al.*, 2001; Dessì *et al.*, 2002).

The soil sample used in the study comes from a sampling section located along the terrace of the river Montevecchio-Sitzerri and consists in the upper part of reddish flotation tailings mud beneath which there is a layer of less contaminated soil and alluvium. The sample was homogenised and divided up so as to obtain representative batches for characterization and decontamination experiments.

Mineralogical analysis revealed that the main phases containing heavy minerals were: siderite, sphalerite and galena, pyrite, cerussite, smithsonite, chalcopirite, anglesite and iron oxides that occur as secondary minerals.

Table **5** shows the particle size distribution of the soils determined by wet screening and the chemical analysis for the major elements. As can be seen, the heavy metals tend to concentrate in the finer fractions.

Table 5: Particle size distribution and chemical composition of the soil samples

Size Class (mm)	Metal Content in Sample A (ppm)					
	Mass (%)	Cu	Pb	Zn	Cd	As
+1	5.5	22	401	582	6	36
-1+0.5	3.4	33	1,138	1,347	11	73
-0.5+0.3	12.3	81	957	1,035	9	50
-0.3+0.1	19.0	79	1,438	1,781	6	59
-0.1+0.074	9.0	85	1,614	1,729	8	50
-0.074+0.038	16.4	95	1,901	2,078	7	48
-0.038+0.020	8.9	103	1,835	2,177	8	39
-0.020	25.5	201	3,341	3,298	15	103
TOTAL	100	110	1,924	2,024	9.3	64

Physical Treatments

Preliminary investigation demonstrated that gravity separation methods alone were not suitable for decontaminating the coarse fractions, probably because the metal species present in whatever form in the soil were not sufficiently liberated. Similarly, classic sulphide mineral flotation techniques do not seem capable of removing metals to an acceptable level, because the surfaces of the minerals contained in the soils are highly oxidized.

A partial decontamination effect has been obtained combining an attrition stage followed by hydrocyclone de-sliming and gravity separation of the de-slimed product. Fig. **2** shows schematically the treatment procedure adopted.

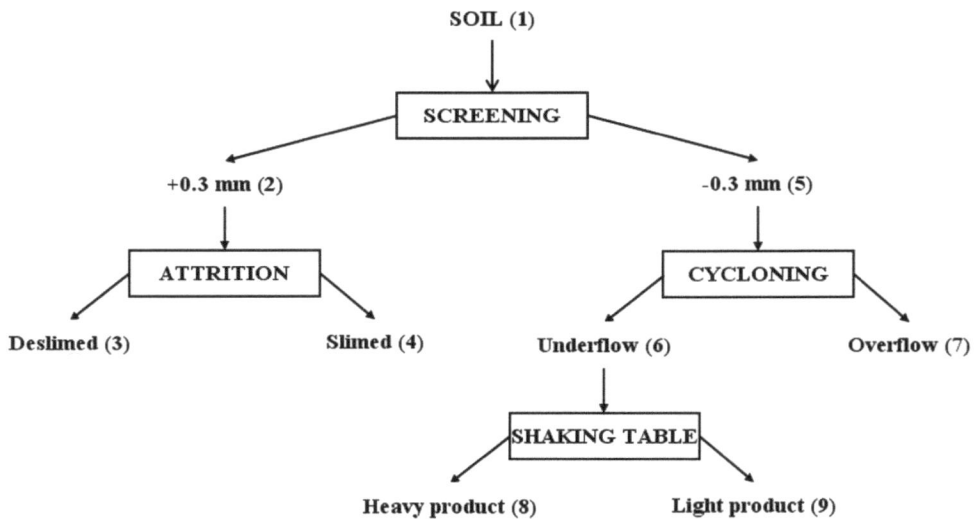

SOIL (1)

SCREENING

+0.3 mm (2)   -0.3 mm (5)

ATTRITION   CYCLONING

Deslimed (3)   Slimed (4)   Underflow (6)   Overflow (7)

SHAKING TABLE

Heavy product (8)   Light product (9)

Figure 2: Processing flowsheet.

Table **6** shows the mass balance and metal content of the different products. The "de-slimed attrition product" and "light product from the shaking table", indicated in bold in the table, yield the mass fraction with low metal concentrations, whereas the "attrition sludge", "cyclone overflow" and "heavy product from the shaking table" yield the mass fraction to be disposed of in the form of toxic and harmful waste. As can be observed, the experiments allowed recovering an overall mass of a little less than two thirds of the original soil mass.

Leaching

Laboratory experiments have been conducted in mechanically agitated reactors, using citric acid and acetic acid solutions and brine of hydrochloric acid and calcium chloride (Dessì *et al.*, 2000; Dessì *et al.*, 2002). The tests showed the brine to be the most effective for removing metals from the soils. Based on the findings of the investigations, the possibility of decontamination by heap leaching has been simulated in the laboratory using the column technique. Column tests have been performed at three scales of 0.6, 1.3 and 3.0 m on soil samples with different degree of contamination (Dessì *et al.*, 2002). Referring to the literature concerning the results achieved with 0.6 and 1.3 m columns, by way of example, only the investigation carried out using a 3 m column is here summarized. Leaching was performed in 3.0 m high, 160 mm diameter plexiglass column. The material to be leached was kept in place at the bottom of the column by means of a filter system, through which the leach solution percolated. The procedure reported below has been adopted.

Table 6: Sample A. Mass balance and metal content of the different products obtained following the processing flowsheet of Fig. **2**

No.	Products	Mass (%)	Metal Content (ppm)				
			Cu	**Pb**	**Zn**	**Cd**	**As**
3	**Deslimed attrition product**	**16.92**	**26**	**387**	**481**	**7**	**48**
4	Attrition slimes –53 µm	4.19	188	2923	3190	17	98
2	Oversize +0.3 mm	21.11	58	890	1018	9	58
9	**Light product shaking table**	**48.40**	**64**	**1165**	**1443**	**5**	**48**
8	Heavy product shaking table	4.10	362	6998	5248	30	117
6	Cyclone underflow	52.50	87	1620	1740	7	53
7	Cyclone overflow	26.39	218	3670	3619	14	107
5	Undersize –0.3 mm	78.89	131	2306	2369	9	71
1	Feed	100.00	116	2007	2084	10	69

-   Agglomeration:

    The soil sample, about 53 kg, has been agglomerated with 2% by weight of carboxymethylcellulose (CMC) and 15% by weight of HCl (0,2M) and $CaCl_2$ (4M Cl⁻) brine, prepared using tap water.

-   Leaching:

    The leach column was filled with agglomerated soil and the material was left to age for 48 hours before starting the tests. Brine of the same composition used for agglomeration was pumped with a peristaltic pump to the top of the column. Percolation problems were recorded in the column. A substantial decrease of percolation from the starting flowrate of 0.22 l/h/m^2 was observed during days of low ambient temperature, *i.e.* close to 0 °C. For this reason the feeding was suspended and the column was allowed "to rest" for 19 days. This temporary suspension of the leaching process had a beneficial effect on percolation ability and thereafter it was possible to continue and complete the treatment with the initial feeding rate. The brine was not recycled in the leaching operations. The leachate, collected in a container at the base of the column was weighed, sampled and the pH, conductivity, Pb, Zn, Cd, and Cu concentrations were measured. Tests were conducted on some of the eluates to precipitate the metals contained therein, varying the pH by means of additions of $Ca(OH)_2$.

- After 90 days leaching, material inside the column was washed with tap water. The eluate was weighed and sampled and the same determinations done as for leaching.

- Finally, the material removed from the leach column was divided up into six slices (slices are numbered 1 – 6 from the bottom to the top). Each slice was weighed and oven dried at 45 °C, pH was measured, relative moisture content calculated and a chemical analysis was performed. Table **7** shows the weight, pH, moisture content and metal content from the top to the bottom of the column after leaching and washing.

The results of the study indicate that the leaching process can be very effective in removing the metal contaminants from the soil. The extractions achieved in the column of 3.0 m were more than 85% for Pb, 70% for Zn and 89% for Cd and the residual concentrations were below existing soil quality guidelines. However, due to the fine texture of the soil and the difficulty of the percolation, better performance has been achieved in the first 1.5 m of the column where residual metal concentration in the leached material satisfies current Italian regulatory limits for "sites for commercial and industrial use", (D.Lgs. n. 152, 3 April 2006, Table **1B** (D.Lgs. 152/06).

Table 7: Cumulative results *vs.* column depth

N.	Depth (m)	pH	Moisture Content (%)	Mass (%)	Metal Content (ppm)			
					Pb	Zn	Cd	Cd
6	0.0-0.4	6.08	27.73	16.68	387	545	2.5	132
6+5	0.0-1.0	5.23	28.17	37.63	959	749	2.3	207
6+5+4	0.0-1.5	5.01	29.11	52.76	997	830	2.5	232
6+5+4+3	0.0-2.0	4.84	29.66	69.64	1,116	873	2.6	239
6+5+4+3+2	0.0-2.4	4.76	29.39	83.22	1,518	932	2.7	256
6+5+4+3+2+1	0.0-2.9	4.66	30.24	100.00	1,795	1,017	2.7	272
Feed					12,245	3,366	25	444
D.Lgs. 152/06					1,000	1,500	15	600

GRANITE WASTE

One pressing problem facing the stone industry, from both a management and environmental impact standpoint, are the increasing amounts of waste accumulating near granite quarries and processing plants. In spite of the sophisticated exploration techniques used nowadays to optimize stone extraction, it is not possible to reduce the amount of waste produced beyond a certain limit. One potential use of granite waste is as graded aggregates for concrete manufacture or as railway ballast, as granulates for road bases or as foundation layers in civil engineering works.

Though a ready market does exist for these products, the cost of long distance transport of low value material such as granite waste does not make this an economically feasible option. Thus additional amounts of granite waste might be disposed off through the production of feldspar and quartz concentrates to be used in the glass and ceramics industries (Ghiani *et al.*, 1996; Ghiani *et al.*, 1997; Ghiani *et al.*, 2007). The technical feasibility of this form of disposal is examined herein.

The investigation was conducted on waste generated by extraction of granite widely occurring in northern-central Sardinia and known as "Sardinian Grey".

The sample was supplied in the form of pieces having maximum size of around 400 mm, and satisfactory liberation was obtained at about 2 mm for potassium feldspar and for plagioclases, whereas finer grinding to below 1 mm was required to liberate the mica.

Comparison of the data from microscopic and chemical analysis made it possible to establish the following mineralogical composition: 27% potassium feldspar, 36% plagioclase, 33% quartz and 4% mica.

The experimental investigation of the possibility of beneficiating the granite waste was conducted on material crushed first in a jaw crusher and then in closed circuit in a roll mill to below 1 mm. The sample thus obtained was used in experiments to assess the efficiency of separation by means of magnetic techniques alone and combined with flotation techniques.

The possibility of removing the magnetic femic components by means of dry magnetic separation has been assessed, after de-dusting to 0.05 mm, using a permanent magnetic roll separator (Permroll) with 1.6 Tesla field intensity. Flotation tests were carried out using a 2.5 l Denver laboratory cell.

Of the different possible combinations, it is proposed a flow sheet that envisages:

- Comminution of raw material to below 1 mm;

- De-dusting of the resulting product at 0.05 mm;

- Magnetic separation of the de-dusted material;

- Further wet grinding of the non magnetic product to below 0.3 mm;

- De-sliming at 0.04 mm of the ground material;

- Selective flotation of the de-slimed material to separate the feldspar minerals from the quartz, following the classic technique using cationic collectors (Amine MFA10, AkzoChemie) in an acid pulp HF.

Table **8** shows the results of magnetic separation combined with selective flotation of granite waste.

Table 8: Results of magnetic separation combined with selective flotation of granite wastes

Type of Granite	Product	Weight (%)	Content (%)					
			K₂O	Na₂O	CaO	SiO₂	Fe₂O₃	TiO₂
Sardinian grey	Magnetic	15.3	5.2	3.2	1.0	64.8	7.4	0.45
	Float	45.0	6.7	5.8	1.4	67.9	0.08	0.01
	Non float	23.0	0.1	0.1	0.1	99.0	0.01	0.01
	Slimes -0.04 mm	16.7	5.2	4.9	1.2	71.0	0.5	0.03
	Total	100.0	4.7	4.0	1.0	75.1	1.26	0.08

The suitability of the feldspar concentrates obtained *via* magnetic separation and/or flotation for use in ceramic mixes, which are prepared for manufacturing glazed porcelain tiles or ceramic underglazes, was assessed by determining linear shrinkage and water absorption at different firing temperatures.

Furthermore, the behaviour of the feldspathic products was studied by varying the proportions of the products obtained by magnetic separation in the formulation of ceramic mixes suitable for the production of porcelain stoneware tile and using the feldspar concentrate obtained by the combination of magnetic separation and flotation techniques in the preparation of engobes, suitable for use as underglazes, as well as in the preparation of glaze compositions (Ghiani *et al.*, 2007). The results clearly demonstrate the possibility of using the obtained feldspathic products in various sectors of the ceramic industry (Ghiani *et al.*, 2007).The laboratory experiments demonstrated that it is possible to use granite waste as a raw material for producing feldspar concentrates for the ceramics industry, thus avoiding its disposal. A relatively simple dry magnetic separation allows for satisfactorily reduction of the Fe content of the waste: the finer the grinding, the greater the removal efficiency achieved. The results show that it is possible to obtain a feldspathic material suitable for use in the production of glazed porcelain tiles. To obtain products that meet more stringent quality specifications, magnetic separation will have to be combined with selective flotation techniques.

In any case, the magnetic separation/flotation combination can be carried out in a flexible manner in relation to both the desired product mix and the product quality characteristics. An intermediate solution might be, for example, classification of the non magnetic product from the preliminary magnetic separation to sizes in the order of 0.3 mm, subjecting the undersize to flotation. The flexibility of this procedure could make it possible to produce simultaneously, depending on market requirements, feldspar materials suitable for manufacturing glazed porcelain tiles as well as materials suitable for the production of higher quality products, in addition to quartz materials to be used in the glass industry.

CONTAMINATED SEDIMENT DREDGING

The investigation aimed at verifying the possibility of applying soil-washing techniques, developed in mineral processing, to eliminate or to reduce the toxic elements found in marine sediments coming from the Porto Vesme area, located in southern-eastern Sardinia (De Gioannis *et al.*, 2009). The collected sample consisted of a coarse size fraction (> 5 mm) mainly composed of vegetable residues (seaweeds, *etc.*) and of organic carbonate (shells, *etc.*) and of a finer size fraction mainly consisting of sand, sediments and lime. Size distribution of the as received sample and metal analysis of the size fraction less than 5 mm are given in Table **9**.

Table 9: Particle size distribution and chemical composition of the sample

Size Class (mm)	Mass (%)	Cd (mg/kg)	Cu (mg/kg)	Ni (mg/kg)	Pb (mg/kg)	Zn (mg/kg)
+5	2.00					
-5+0.5	1.63	46.0	100.0	48.0	1,799.5	4,498.7
-0.5	96.37	24.9	69.2	45.8	863.2	4,558.1
Total	100.00	25.3	69.8	45.8	878.8	4,557.1

Table **9** shows that most of the material, more than 96%, has a size below 0.5 mm, where the heavy metal content exceeds the limit established by current Italian legislation in areas to be used for commercial and industrial activities (D.Lgs. 152/06, Table 1B). The size class up to 0.5 mm mainly consists of vegetable residue and of calcareous materials of organic origin (shells, *etc.*) with a small fraction of lithics and quartz sand. Furthermore, the chemical analysis shows a non- negligible heavy metal content which is most likely present as fine particles inside of the bioclasts and mixed with vegetable matter. Considering that this size fraction corresponds to less than 4% of the sample, it was decided to exclude it from cleaning treatment. On the basis of the characteristics of the material the following procedures were adopted:

- Size classification by means of a sieve in order to reject the coarser fraction up to 5 mm (vegetable residuals and calcareous materials of organic origin);

- Further reject of the size fraction up to 0.5 mm always by means of sieving;

- Hydrocyclone desliming of the 0.5 mm fraction in order to obtain an underflow suitable to be treated with gravimetric techniques, such as shaking tables or spirals, and to remove the contaminated overflow ($d_{80} < 15$ μm);

- Gravity processing of the hydrocyclone underflow on a shaking table aiming at verifying the possibility of removing heavy fractions characterized by contaminant contents and of obtaining a light fraction with low metal concentration, below the limits established by current Italian legislation (D.Lgs. 152/06; Decreto Ministeriale n. 186, 5 April 2006).

The results (Table **10**) of the hydrocyclone de-sliming demonstrated a reduction of heavy metal content in the underflow but not enough to fall within the limit established by current Italian legislation (D.Lgs. 152/06, Table 1B, areas to be used for commercial and industrial activities).

However, with the hydrocyclone desliming it is possible to remove an overflow product, about 10% of the feed weight, with $d_{80} = 15$ μm and with a large heavy metal concentration, which is hard to clean with classical gravity techniques. Gravity processing of the hydrocyclone underflow on a shaking table allows obtaining a "light" product, about 80% by weight of the feed, with a heavy metal content below the limit established by current Italian legislation in areas to be used for commercial and industrial activities (D.Lgs. 152/06, Table 1B). Moreover, the leaching test carried out according to UNI 10802 standard methods on the same light fraction showed that concentrations of all analysed elements are below the limits established by current Italian legislation (Decreto Ministeriale n. 186, 5 April 2006) for inert waste reuse. Similar positive results can be expected also by performing gravity separation using equipment characterized by higher treatment capacity (*i.e.* spirals). Since the metals in the sediments are mainly present as heavy minerals, probably unintentionally discharged into the harbour waters in loading/unloading merchant ships, remarkable results were obtained applying a mechanical washing treatment, characterised, in particular, by a gravimetric stage.

Table 10: Results of sediments dredging treatment

	Product	Weight (%)	Content (mg/kg)				
			Cd	Cu	Ni	Pb	Zn
	+5 mm	2.00					
	-0 +0.5 mm	1.63	46.0	100.0	48.0	1,799.5	4498.7
	Sub-total	3.63					
Hydrocyclone	Overflow	9.77	45.9	139.5	146.0	2,073.8	8,285.4
	Underflow	86.60	22.6	61.3	34.5	726.5	4,137.4
	Sub-total	96.37	24.9	69.2	45.8	863.2	4,558.1
Shaking table	Heavy	6.10	171.8	341.4	97.2	3,883.7	40,829
	Light	80.50	11.3	40.1	29.7	487.2	1,355.9
	Sub total	86.60	22.6	61.3	34.5	726.5	4,137.4
	Total	100.00	25.3	69.8	45.8	878.8	4,557.1

In fact, after the mechanical washing treatment, the contaminants were concentrated in the heavy products (only a fifth of the total mass), whilst most of the material (80% by weight) showed heavy metal concentrations lower than the limits adopted by the Italian legislation for soils located in industrial areas, hence fulfilling the regulation limits foreseen for reuse of sediments in infrastructural confined disposal facilities. Therefore, the residue resulting from the dredging activities was successfully treated by relatively simple and low cost mechanical phases.

FLY ASH FROM COAL POWER STATION

Very large quantities of fly-ash are currently produced in coal-fired power stations all over the world, giving rise to major technical problems and additional financial burden. Therefore, the re-use of this kind of material for civil engineering applications would offer significant advantages, dispensing with the need and cost of disposal in suitable landfill systems and providing in addition some return on the raw-material market. Currently, fly ash is being used to an ever increasing extent in building and construction applications (cement additive, lightweight aggregate).

Another important application is to use unburned carbon as filter media, in place of activated carbon or as a substitute for industrial carbon. In fact, the unburned coal particles are structurally modified while passing through the combustor resulting in the formation of macro-porous surfaces with presence of silica and alumina compounds, so that, after separation from the true ash, they may constitute a suitable precursor for the production of bulk adsorbents. However the most common application is actually the use of fly-ash as a component of concrete. Tests carried out by different organisations showed that carbon could be a deleterious substance in concrete because it adversely affects air entraining admixtures commonly used to impart desirable properties to concrete, including increased durability.

International standards have established a 5% maximum content of unburned coal in fly ash usable for cement application, although some concrete producers set the maximum allowable Loss on Ignition (LOI) value at less than 4%.

In order to reduce and prevent the landfilling prone to increasing restrictions, as well as to increase the use and value of fly ash with respect to various applications, a beneficiation process becomes necessary. Regarding the choice of the technology, the main requirement is that it must be suitable for the treatment of very finely divided matter, taking into account that fly ash recovered in electrostatic precipitators is usually below 150 μm, with a considerable proportion finer than 50 μm.

A second and not less important requirement is the low-cost of a treatment process, due to the fact that fly ash has a low commercial value especially for the most common applications.

In the last years, the IGAG (Environmental Geology and Geoengineering Institute of CNR, UOS of Cagliari, Italy) together with DIGITA (Department of Geoengineering and Environmental technologies University of Cagliari, Italy) have carried out an investigation aimed at verifying the possibility of beneficiating fly ash to obtain a product suitable for cement manufacturing or other advanced applications and a low-heat fuel to be recycled back to the boiler or to use in immobilization techniques (Ciccu *et al.*, 1997; Ciccu *et al.*, 1999; Ciccu *et al.*, 2000; Ciccu *et al.*, 2005).

The proposal which includes a size classification for recovering the finest fraction that already contains little combustible matter and eliminating unburned carbon from the coarser fraction by means of electric separation is here reported.

A sample of fly ash from a coal power station located in the south of Italy has been tested. Microscopic investigation showed that the non-combustible fraction of the fly ash consists mainly of isolated particles of spherical shape, whereas unburned carbon is found as elements of irregular shape, often in aggregates. Diffractometric analysis revealed the presence of mullite and of a phase similar to graphite that may be the product of coking. Magnetite is also present as the final step of pyrite oxidation. The larger part of the fly ash consists of an amorphous phase with extremely variable chemical composition, as microprobe investigation showed.

Dry size analysis of the sample gave the results shown in the following Table **11**.

Table 11: Size analysis of the fly ash sample

Size Class (μm)	Weight (%)	L.O.I. (%)	L.O.I. Distr.(%)
+45	28.11	15.69	65.94
-45+37	6.45	5.09	4.91
-37	65.44	2.98	29.15
Total	100.00	6.69	100.00

From the results of size analysis, size classification technique appears to be an interesting approach to apply on account of the fact that unburned particles are generally coarser in size than non-combustible matter. On the basis of the characteristics of the material, the following flowsheet was proposed:

- A screening stage to recovery a low-LOI product meeting the requirements for cement applications, the quality of which can be controlled by the separation size depending on the mesh opening and on the setting of the screen;

-   The electrostatic separation of the screen oversize into two products: again a low-LOI fraction to be blended with the sieve fines and a carbon rich fraction to be recycled to the boiler or to be used for other applications.

As fly ash chiefly has been found to consist of isolated particles having prevailingly a spherical shape, whereas unburned carbon occurs as elements with irregular contour, often as aggregates, size separation of the low-carbon fraction must be carried out very gently using apparatuses not involving strong actions on the particles (more particles of unburned matter are likely to break into smaller fragments that would pass in the underflow).

Size classification tests have been carried out using a Vibrowest circular screen with different mesh openings and variable setting of the counterweights in order to control the kind of movement of the fly ash mass flowing through the machine. Using a suitable set-up, capacity and residence time, it was possible to obtain an undersize with a LOI content of 2.98% and a recovery of 68% (Ciccu *et al.*, 1999).

Before undertaking the electrical separation tests, the behaviour of selected samples of unburned particles and of carbon-free ash subject to contact charging was studied by measuring the electric charge per unit mass with a Faraday well connected to an electrometer. Ash is negatively charged by contact on the vibrating chute, while unburned particles are positively charged by conductance on a grounded metal surface while passing through a static field generated by a negative voltage.

The amenability of the screen oversize to be treated with electrostatic separators has been tested using different electrostatic separators:

-   A classical drum separator (Ciccu *et al.*, 1997);

-   A "cone-shaped" prototype, that consists of: a vibrating stainless steel feeding device having also the task of initiating the charging process and of distributing the solid stream onto the carrier surface below; a stainless steel cone-shaped surface, connected to the ground or to a voltage source and revolving along its vertical axis; a set of electrodes for the establishment of a static field, the arrangement of which and the position with respect to the carrier surface can be modified according to needs; a section for the separate collection of the products and for cleaning of the surface; an optional encasement, inside which the above items are hosted, allowing the measurement and the control of temperature and humidity of both the material and the surrounding environment aiming at providing, if necessary, the most favourable conditions for separation; a voltage source (Ciccu *et al.*, 2000);

- A "metal belt" prototype (Peretti *et al.* 1998), that consists of: a feed device for initiating the charging process and discharging the solid stream onto the belt; a metal belt, with adjustable inclination electrically grounded or connected to a voltage source; a set of electrodes for creating a static field, their arrangement and position with respect to the belt can be modified according to requirements; a voltage source (Ciccu *et al.*, 2005).

Referring to the literature about the results achieved with each of the mentioned separators, by way of example, the results obtained considering a stage of size classification using a Vibrowest circular screen followed by electrostatic separation (of the screen oversize product) using a "cone-shaped" prototype are reported in Table **12**.

Table 12: Beneficiation results (Combination of dry screening and electrostatic separation)

Products	Fly Ash Sample		
	Yield (%)	**LOI (%)**	**LOI - Distr.(%)**
Benef. Ash	86.56	3.12	41.01
Middlings	7.52	13.48	15.39
Comb. fraction	5.92	48.50	43.60
Total	100.00	6.59	100.00

The increasingly restrictive regulations foreseen for landfills and the consequent need to reduce the amount of fly ash to be disposed off in this way, makes beneficiation essential for increasing both the use and value of this by-product. Electrostatic separation (applied alone or combined with a stage of size classification) offers a potentially promising solution to this problem as with this technique it is possible to treat the hot, dusty fly ash in the dry state. Results obtained are encouraging and a project is being started with a pilot plant having a convenient throughput capacity.

CONCLUSIONS

Many countries have set up strategic programs about criteria, procedures, regulations for the Industrial Waste Management. Even though legislation and normative may differ from country to country, a common waste management hierarchy - reduction; re-use; recycling and disposal - is generally considered. Remediation of "polluted site" is the final step in implementing the waste hierarchy.

Mineral processing techniques can give a substantial contribution to the problem of Industrial Waste Management. In fact, the steps to follow in order to achieve the result of reducing and recycling waste as well as remediating "polluted site" are generally the same as those applied in mineral processing, in a few words: solid/solid and solid/liquid separation after physical-mechanical or wet liberation.

Nevertheless the necessity of applying more or less complex and onerous flowsheets and the quality of achievable results can strongly differ from case to case as evidenced in soil polluted by mine activities.

Simple (solid/solid) gravity techniques to treat the waste material from Monte Narba mine site allow to obtain a pre-concentrate containing Pb, Ag and $BaSO_4$, suitable for further processing in an ore dressing plant, and a product, about 80% of the starting mass, substantially inert. Otherwise, the remediation of the soils from Montevecchio mine, always contaminated by sulfides minerals, asks for the use of more complex and onerous flow sheets even applying strong leaching techniques, but without to achieve high technical efficiency and useful products.

Furthermore the examples reported here show the technical possibility of obtaining marketable products from granite waste and from fly ash from coal power station at reasonable process costs. Moreover the feasibility of applying soil-washing techniques, developed in mineral processing, to eliminate or to reduce the toxic elements found in contaminated sediment dredging has been evidenced.

However the choice of method, whether *ex situ* or *in situ* (stabilizing and/or confining techniques in particular), depends on a number of parameters related to the type and extent of contamination, future land use of the reclaimed sites and financial resources required.

ACKNOWLEDGEMENT

Declared none.

CONFLICT OF INTEREST

The author(s) confirm that this article content has no conflict of interest.

REFERENCES

Ciccu, R., Ghiani, M., Peretti, R., Serci, A., Zucca, A. (1997). Electrostatic separation of unburnt coal for fly-ash beneficiation. In: *Proceedings of XX International Mineral Processing Congress, Waste Treatment, Recycling and Soil Remediation*, Hoberg H. Ed., GDMB Publisher, Aachen, Germany, 135-146.

Ciccu, R., Ghiani, M., Muntoni, A., Serci, A., Peretti, R., Orsenigo, L.G., Quattroni, G., Zucca, A. (1999). The Italian Approach to the Problem of Fly Ash. In: *Proceedings of the International Ash Utilization Symposium*, Lexington, Kentucky, USA, 591-601.

Ciccu, R., Ghiani, M., Peretti, R., Serci, A., Zucca, A. (2000). A New Electrostatic Separator For Fine Particles. *In: Proceedings of the XXI International Mineral Processing Congress*, Massacci P. Ed., Roma, Italia, Elsevier Publisher, A7-42 – A7-51.

Ciccu, R., Ghiani, M., Peretti, R., Serci, A., Zucca, A. (2005). Development of a new electric belt separator for fine particles. In: *Proceedings of the XI Balkan Mineral Processing Congress, Mineral processing in sustainable development*, Fetahu K. Ed., Albpaper Publisher, Tirane, Abania, 259-266.

Dessì, R., Ghiani, M., Fadda, S., Muntoni, A., Peretti, R., Persod, P., Zucca, A. (1999). Soil Decontamination by Mineral Processing Techniques: The Case of Montevecchio Levante Site (Sardinia, Italy). In: *Proceedings of Rewas'99*, San Sebastian, Spain, vol. III, 2501-2508.

Dessì, R., Fadda, S., Peretti, R., Serci, A., Zucca, A. (2000). Soil decontamination at the Montevecchio-Levante Mine Site with experimental washing and leaching techniques. In: Annali di Chimica, *Società Chimica Italiana*, 90, 687-694.

Dessì, R., Ghiani, M., Peretti, R., Persod, P., Pilurzu, S., Saju, A., Zucca, A. (2001). Caratterizzazione di aree contaminate da attività mineraria: i casi di Montevecchio-Levante, Monte Narba, Sedda Moddizzis, In: *Atti della Facoltà di Ingegneria di Cagliari*, n. XXIX, 44, 5-13

Dessì, R., Zucca, A., Peretti, R., Serci, A., Tambouris, S., Papassiopi, N., Paspaliaris, I. (2002). Evaluation of Leaching Techniques for the Remediation of Montevecchio Soils. In: *Proceedings of Seven International Conference on: Environmental Issues and Waste management in Energy and Mineral production, SWEMP 2002*, Cagliari, Italy, 815-822.

De Carlo, I., Dessì, R., Ghiani, M., Peretti, R., Persod, P., Pilurzu, S., Zucca, A. (2001). Environmental remediation of the Monte Narba mine site in South-East Sardinia. In: *Proceedings of 9th Balkan Mineral Proceessing Congress*, Istanbul, Turkey, 565-570.

De Gioannis, G., Muntoni, A., Peretti, R., Serci, A., Spiga, D., Zucca, A. (2009). Comparison of different approaches to treatment of residues resulting from contaminated sediments dredging. In: *Proceedings Sardinia 2009, Twelfth International Waste Management and Landfill Symposium*, Cossu R., Diaz L.F. e Stegmann R. Eds., CISA Publisher, Italy, Extended Abstracts, 1183-1184.

Ghiani, M., Peretti, R., Zucca, A., Angius, R. (1996). Production of raw materials for the ceramic and glass industry from granite quarrying waste. In: *Proceedings of Fourth International Conference on: Environmental issues and waste management in energy and mineral production, SWEMP'96*, Cagliari, Italy, Ciccu R Ed., 1003-1010.

Ghiani, M., Oi, M., Peretti, R., Zucca, A. (1997). Produzione di materie prime per l'industria ceramica dagli scarti delle cave di granito: il caso del materiale di Sarule. In: *Ceramica Acta*, n. 2-3, 5-17.

Ghiani, M., Serci, A., Peretti, R., Zucca, A., Angius, R. (2007). Utilization of granite exploration waste, In: *Proceedings of the XII Balkan Mineral Processing Congress*, National Technical University of Athens, Athens, Greece, 577-582.

Peretti, R., Ciccu R., Zucca A., Ghiani, M., Serci, A., Borghero P. (1998). *A new belt electrostatic separator for fine particles*", Patent RM 98 A 000156, International code: B 03 C.

Progemisa S.p.A. (1998). *Environmental Characterization of Contaminated Sites*, Report of EU Project ROLCOSMOS, Contract No. BRPRCT96-0297, 1997-2000.

Progemisa S.p.A. (2000). *Progetto di recupero ambientale e valorizzazione del centro minerario di Monte Narba, Comune di San Vito*, Rapporto all'Assessorato dell'Industria della Regione Sardegna.

Send Orders of Reprints at reprints@benthamscience.org

CHAPTER 2

Challenges in the Separation of Plastics from Packaging Waste

M.T. Carvalho*

CERENA, Instituto Superior Técnico, Lisboa, Portugal

Abstract: Packaging waste are commonly used as new raw materials reducing the depletion of natural resources. The separation between the different materials is mandatory if the application in high value added products is desired. Today, one challenge for recycling is the separation of shredded polymers with approximate density. In this chapter the specific problems related with the separation of PS (Polystyrene) from PET (Polyethylene Terephthalate) and PVC (Polyvinyl Chloride) are addressed.

The most important international literature describing experimental studies are reviewed and a summation of the experimental work supervised by the author is presented. The application of gravity concentration and froth flotation is proposed. The results of the experimentation carried out with representative samples of post-consumer packaging plastics at laboratory scale and in a pilot plant continuously operated are presented. The study confirmed the process feasibility but highlighted some difficulties and limitations.

Keywords: Recycling, Froth flotation, Gravity concentration, PET, PVC, PS.

INTRODUCTION

All over the world, today, there is a great effort in the recycling of the materials existing in waste aiming at the saving of natural resources and preventing the building up of landfills. Generally, due to the differences in the characteristics of post-consumer items like composition, size, presence of hazardous materials, *etc.*, they are classified in origin based categories such as "End of Life Vehicles", "Electric and Electronic Waste" and "Packaging Waste", among many other. Commonly, each type of waste is collected, transported and processed separately from the other.

For the managing of packaging waste, that is the object of the present work, in Portugal, as in several countries, there is a "Green Dot System", Sociedade Ponto Verde (SPV), a private, non-profit company, that promotes the selective

---

*Address correspondence to M.T. Carvalho: CERENA, Instituto Superior Técnico, Av. Rovisco Pais, 1049-001 Lisboa, Portugal; Tel.: +351218417425; Fax: +351218417389; E-mail: teresa.carvalho@ist.utl.pt.

collection, recovery and recycling of packaging waste at national level. The materials that are recycled are wood, glass, metals, paper & board and plastics, being metals and plastics deposited by citizens in the same place (yellow bin). In 2009, 286,488 t of household packaging waste, of which 62,015 t of plastic, were collected in Portugal. Plastic and metal packages are deposited together in the same bin (in Portugal these are yellow bins). The contents of the bins are transported to sorting plants where the packages made with different materials are separated to be sent to the respective recycling industry. Table **1** shows a typical composition of the yellow bin stream. This resulted from a long term sampling study (2 years) carried out under a research study funded by SPV (Carvalho *et al.*, 2007a). The sampling was made in the feed stream of a sorting plant (Tratolixo company), located 35 km from Lisbon. It should be highlighted that the composition of waste streams is heterogeneous and geographical and time varying. In the case illustrated in Table **1** the sampling was carried out in years 2007 and 2008 in a predominantly urban area.

Table 1: Composition of the yellow bin stream received by Tratolixo sorting plant

	Plastics	**Metals**	**P&C[1]**	**PLA[2]**	**Glass**	**Other**
% Weight	49.8	13.3	12.3	9.3	3.9	11.3

[1]P&C: Paper and cardboard, [2] PLA: liquid food carton.

It should be pointed out that, as it can be observed in Table **1**, although only plastics and metals should be deposited in the yellow bin, due to the lack of information or environmental education, almost 30% of the composition correspond to other materials.

The waste from the yellow bin is transported to sorting plants where the materials are separated. In modern plants, PLA and P&C are separated by ballistic separators, metals by magnetic and electromagnetic (based in eddy currents) separation and plastic film is separated by aspiration. Although these operations are automatic, manual sorting is always present. Household plastic packages are mostly made of PE (Polyethylene), PET and PP (Polypropylene), although PS and PVC occur too. Most polymers cannot be recycled together due to their incompatibility during melt processing. So, they should be separated into the individual plastics if they are to be used in high value recycling products. This is done either by manual and/or optical sorting. Each type of plastic is then sent to the corresponding recycling industry where it is subjected to additional sorting processes.

Manual and optical sorting are quite efficient techniques to separate large objects. However, commonly, the objects are grinded for materials separation. This is what happens, for instance, in the PET recycling industry, where the bottles, made with PET, have to be grinded to some millimeters for materials liberation, because caps and labels are often made with other polymers. Below approximately 10 mm manual and optical sorting become inapplicable or the loss in efficiency is high. Rigid PE and PP, after grinding, can be separated from the other polymers by static flotation in water, contrarily to what happens with the other polymers, because they have a lower specific weight than water. This process is not challenging as long as the formation of air bubbles is avoided since, due to the polymer's high hydrophobicity, the air bubbles concentrate at the polymer particle surface making it float, even when its specific weight is higher than 1 g/cm^3. Packaging PET and PVC have similar density, respectively, 1.34-1.39 and 1.35-1.45 g/cm^3, while the density of PS is only slightly lower, 1.05-1.10 g/cm^3 (Billmeyer, 1984). So, the separation of these three plastics, when granulated, is a challenging task, being a topic of research today.

SEPARATION OF PS FROM PET AND PVC

Gravity concentration is commonly used in mineral and coal processing to separate minerals which have a minimum density difference. The suitability of the application of gravity concentration processes to the separation of a particular set of minerals is generally evaluated by the use of a parameter CC (1), the "concentration criterion" (Wills, 1998).

$$CC=(d_h-d_f)/(d_l-d_f) \tag{1}$$

where d_h, d_l and d_f are, respectively, the specific gravity of the "heavy" and "light" species, and fluid. The fluid, usually water, can also be air, another liquid or fluid, or even a suspension of solid particles in water.

According to this criterion, the separation should be easy when CC > 2.5, for a particle size down to 75 μm, but impossible at any size when CC < 1.25.

The application of this criterion to mixtures of PS, PET and PVC, by direct use of the tabled densities of these polymers, shows that the separation between PS and PET and between PS and PVC should be easy (3.4 < CC < 7.8 and 3.5 < CC < 9, respectively) and impossible in the case of the separation of PET from PVC (1.03 < CC < 1.32).

Carvalho *et al.,* (2007b and 2009) carried out experimentation with samples of virgin and post-consumer packaging using wet shaking table and hydraulic bed classification which are gravity concentration processes. These authors showed that the separation of PS from mixtures composed by PS, PET and PVC, in fact, is not as easy as the concentration criterion would predict. Furthermore, unexpectedly, using a wet shaking table, these authors could produce an almost pure PET product from mixtures of PET, PVC and PS. Nevertheless, these observations agree with the conclusions of Pascoe and Hou (1999) who, using a Larcodems separator (a gravity concentration equipment) to separate PVC and PET, showed that the particle thickness and pre-conditioning of samples with reagents have an important role in the plastics behavior.

When using "real" samples of post-consumer plastics the problem becomes more complex. Firstly, it should be pointed out that, as referred above, the density of one type of plastic is not constant, varying within limits, with the type and concentration of the additives that are present. Moreover, the additives can change the chemical properties and, consequently, the surface chemical properties, with possible modification of surface wettability. As well, the particle "history", namely contact with contaminants, can play an important role in the separation because the contaminants can affect, as well, the physical and chemical properties of the surface namely surface wettability. Finally, even when coming from the same object, two particles although within the same size fraction can have different shape. For example, the thickness of a bottle may not be uniform.

Carvalho *et al.,* (2009) used an elutriator to process samples of post-consumer mixtures of PET, PVC and PS. Among the multiple separation methods and devices used in classification and gravity concentration, the hydraulic or fluidized bed classifiers is the simplest and theoretically most efficient one. It is a low cost process and the equipment is easy to operate and control being operationally robust. The settling of particles against an upward flow depends on the particle size, velocity of the pulp, overflow volume fraction of solids, overflow and underflow pulp densities and pulp viscosity.

In the referred work it was observed that, although using close size intervals (2-4 mm and 4-5.6 mm), contrarily to what should be expected taking into consideration the "concentration criterion", the separation was not "easy" nor perfect. There was PS contamination in the denser (underflow) product while a significant percentage of PET and PVC appeared in the overflow product. When the upward velocity of water was higher, the overflow was more contaminated with PET and PVC while PS appeared in higher proportion in the underflow product when the upward velocity of water was lower.

The pre-conditioning of the sample with calcium lignosulphonate (CaLS), a reagent commonly used in froth flotation, previously to gravity concentration, decreased substantially the contamination of the overflow product, although a small percentage of PS appeared in the underflow product, mainly when the finer particle sample was used. Table **2** presents a summary of the results obtained when pre-conditioning with CaLS was used.

Based on the above, it should be concluded that the "concentration criterion" can be used but only as a guideline in the case of polymers because, besides specific gravity and particle size and shape, other factors such as particle surface properties can have strong effects in the separation.

Table 2: Composition of the fluidizing bed separation PS product (overflow) Samples were pre-conditioned with CaLS

		Size range (mm)	PET (%)	PVC (%)	PS (%)
Feed		2-4	88.18	8.41	2.91
		4-5.6	91.14	6.62	1.63
Overflow (%)	Grade	2-4	21.72	3.33	74.72
		4-5.6	28.79	3.57	67.52
	Recovery	2-4	0.22	0.24	54.50
		4-5.6	0.29	0.53	83.22

SEPARATION OF PET FROM PVC BY FROTH FLOTATION

Similarly to the flotation of minerals, froth flotation applied to the plastic flotation is based on the selective modification of the polymers particle surface by creating an adequate physico-chemical environment that lead to physical separation of the different polymers. The separation between different plastic types may be achieved if the wettability of the plastic surface can be selectively modified. Although many mechanisms have been proposed to explain the selective separation of polymers based on their surface modification, the knowledge today is still incomplete.

Shen *et al.,* (1999) made an extensive review on the proposed mechanisms to date to explain the selective plastics wetting for mixtures of PET and PVC. These were: 1) gamma flotation, where the selective flotation was achieved by the reduction of the liquid surface tension to a value between the critical surface tension of the two plastics; 2) selective wetting of the plastics by selective surface chemical modification or by adsorption of surfactants and 3) selective wetting by physical modification or surface oxidation using a physical method such as plasma treatment and corona discharge. Fraunholcz, in 2004, updated the knowledge on this subject in a first part of a review paper (Fraunholcz, 2004).

CASE STUDIES

For what concerns experimentation, some very important works have to be highlighted. Drelich *et al.,* (1998) applied an alkaline treatment followed by froth flotation. The alkaline treatment, according to these authors, promotes the hydrolysis of the PET surface, increasing the number of hydrophilic groups (hydroxyl and carboxyl) and surface roughness. These authors used 1-3%wt. of NaOH, during 15-30 minutes, at 70-80 °C, to reduce substantially the hydrophobicty of PET while PVC particles were only slightly affected. Marques and Tenório (2000) and Pongstabodee *et al.,* (2008) used a different approach. They did not use any pretreatment before flotation but used surfactants (CaLS in both cases) in a strong alkaline solution (pH 11 and pH 12). Marques and Tenório observed that an hour of conditioning improved the results while Pongstabodee *et al.,* achieved better results when an electrolyte ($CaCl_2$) was used. Fraunholcz *et al.,* (2000) showed that a combination of an alkaline treatment followed by surfactant adsorption (quebracho and arabic gum were used) resulted in a synergetic effect with improved separation and lower reagent consumption. Burat *et al.,* (2009) used an alkaline treatment to render both PET and PVC hydrophilic and then a plasticizer (diethylene glycol dibenzoate) to increase the hydrophobicity of PVC.

For what concerns surfactants, lignosulphonates are the most commonly referred reagents used in the froth flotation applied to PET and PVC. These are anionic reagents that, according to Le Guern *et al.,* (2000), are mainly adsorbed electrostatically on PET and PVC surfaces which, for these polymers, may be considered as negatively charged when suspended in water at neutral pH, so that the Ca^{2+} cations in the solution bridge the lignosulphonate moiety to the plastic surface.

It should be pointed out that the majority of the experimental studies reported in the literature were carried out with virgin or artificial plastic samples, most using polymers with different colors to facilitate the subsequent analysis of products. However, it should be remembered that the same polymer, depending on the additives, which are related with the application, may have different surface physical and chemical properties. PVC from pipes, for instance, has different properties than PVC from packaging. For these reasons, most data reported in literature should be considered as an indication of the expected behavior when using post-consumer plastics.

Froth Flotation Applied to Post-Consumer Plastics

Carvalho *et al.,* (2010) applied the combined treatment, as described by Fraunholcz (2000), to representative samples of post-consumer packaging plastics (see Fig. **1**).

In the alkaline treatment, they tested two caustic reagents, NaOH and the detergent commonly used in the plastics washing from organic contaminations by Selenis Ambiente (today Avertis), a PET recycling Portuguese company. This detergent has a proprietary composition developed by the company. It is known that it has NaOH in its composition. The alkaline treatment was applied at 90 °C. They tested two surfactants, CaLS and Hostaphat. This reagent, that has a significant effect in PET's wettability, is an etanolic solution of alkane phosphonic acid being a component of the above mentioned detergent.

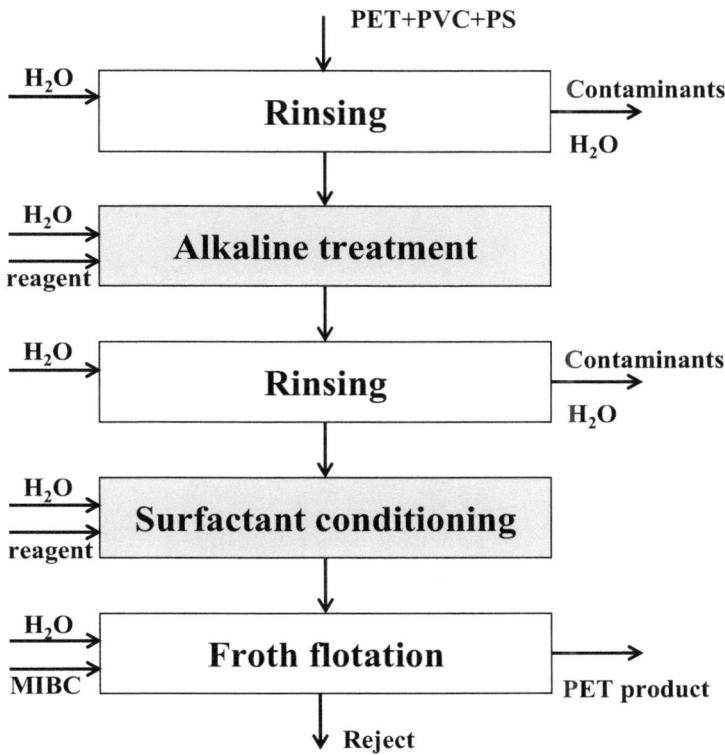

Figure 1: Schematic diagram used in the bench scale tests.

The authors concluded that the commonly used industrial process of plastics washing with detergent is, in fact, an alkaline treatment, with significant effect in the wettability of PET, as long as the froth flotation is performed short after the washing process.

In the referred study, the combination of reagents/treatment that led to the best results was NaOH solution, in the alkaline treatment, and Hostaphat, as surfactant. Using a representative sample of household packaging, collected in the yellow bin stream, with 91.2% PET, 6.5% PVC and 1.8% PS, with particle size 2-4 mm, they achieved a non-floated product formed by 98.9, 0.59 and 0.14% grade, respectively, in PET, PVC and PS. 97.0% of the PET present in feed was recovered in the non-floated product while 95.9 and 90.7% of the PVC and PS, respectively, were recovered in the floated product. This was considered to be satisfactory considering that only one stage of flotation was used.

CONTINUOUS OPERATION

As in mineral processing, after satisfactory results are obtained at bench scale experimentation, before industrial transposition takes place, pilot testing in a continuously operated plant must be carried out. Carvalho *et al.,* (2010) adapted a mineral processing pilot plant (Fig. **2**), owned by a Portuguese mining company to be used with plastics.

Figure 2: A detail of Somincor's pilot plant.

A representative sample of post-consumer packaging plastics from the yellow bin was collected. All the objects made with other materials except PET, PVC and PS were removed. Then, the following standard industrial procedure was carried out: 1) washing, at 90 °C, with detergent; 2) shredding in a cutting shredder equipped with a screen with 12 mm aperture; 3) removal by screening of the -3 mm particles, consisting mainly of paper and glue; and 4) removal of PE and PP by sink-float separation in water. The resultant product, the 3-12 mm size fraction, weighing more than 400 kg, was used in experimentation.

The sample, composed of 85% PET, 2.5% PVC and 11.9% PS was subjected to a combination of alkaline treatment (NaOH and detergent were tested) and surfactant adsorption (CaLS and Hostaphat) followed by froth flotation according to Fig. **3**. The operating conditions used in pilot scale experimentation, namely chemical reagents concentration, temperature and residence times, were transposed from bench scale tests as referred in Carvalho *et al.,* (2010).

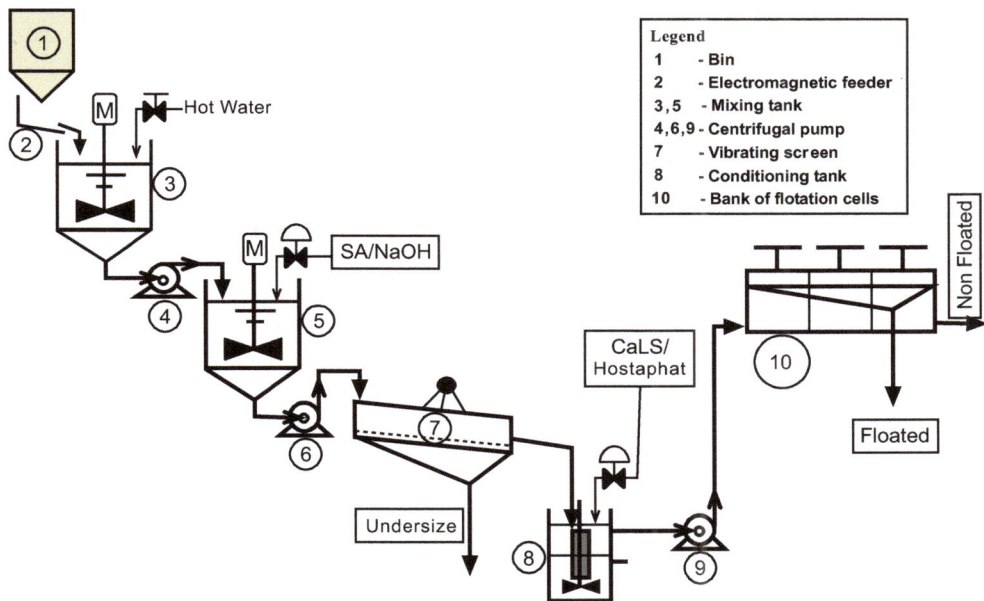

Figure 3: Flow sheet used in the pilot experimentation.

The authors found that, contrarily to what happened in the bench scale experimentation, in pilot tests, the best results were achieved with the alkaline treatment made with the detergent and using CaLS as surfactant. The main cause for this was considered to be the temperature during alkaline treatment, which was lower during pilot experimentation than in bench tests. This agrees with Drelich *et al.,* (1998) who showed that temperature, along with alkaline solution concentration, has a significant effect in the recovery of PET and PVC in the flotation products and therefore in the selectivity of the flotation. The same effect was observed in PS flotation.

A significant upgrading in PET was obtained in a single bank of mechanical flotation cells. The concentrate, containing 83% of the PET in feed, attained

97.2% PET, 1.1% PVC and 1.1% PS in grade. However, with only one stage of flotation it was not possible to attain a PET product in accordance with the Portuguese recycling industry specifications which stipulate that the PVC grade in the PET product should be less than 0.2%. In these conditions, a significant amount of PET would also be lost in the floated product.

As in mineral processing field, the achievement of high purity products from post-consumer plastic packaging demands the use of more or less complex circuits, with scavenging and cleaning stages.

Finally, it should be highlighted that the main limitation to the application of mineral processing based techniques, namely froth flotation, to shredded post-consumer plastic, mainly in continuously driven operation, is the lack of online/on stream methods for the analysis of the composition of separation products. In the experimentation undertook, the evaluation of products composition was made by a selective dissolution process (Santos *et al.*, 2007). This is an off line technique that takes at least several hours. The results from experimentation are known with an unacceptable delay, especially in optimization studies in continuous operation since the acquisition of results in time is critical to know the responses to variable manipulation.

CONCLUSIONS

Although some decades have passed since the first studies on the application of mineral processing based methods to plastic separation this area of knowledge is still challenging. For what concerns the separation of rigid packaging plastics with density lower than water's (the case of PP and PE) from the other plastics, the process is industrially carried out today efficiently by separation in water. Commonly, reagents are used to prevent the formation of air bubbles and to improve higher density plastics wettability. The remaining types of plastic are mainly PET, PVC and PS. The separation of PS, which has a slightly lower density than the other polymers, is possible using gravity concentration. The efficiency of the separation improves when the pulp is pre-conditioned with a convenient surfactant like CaLS.

The PVC commonly used in packaging has a density close to PET's. The separation between these two polymers can be efficiently achieved by froth flotation. The present state of the art shows promising results obtained with this technique, indicating its potential for separation of plastics from post-consumer waste that are otherwise difficult to separate. It is sure that the pulp containing the shredded mixture of plastics must be subjected to a caustic environment before froth flotation takes place to increase PET's wettability. The pre-conditioning with a convenient surfactant like CaLS improves the process efficiency. This has been shown in a continuously driven pilot plant fed with real post-consumer packaging plastics. Nevertheless, the main challenge today is the availability of on line procedure for products quality evaluation.

ACKNOWLEDGEMENTS

The author acknowledge Sociedade Ponto Verde for the financial support of the study included in SEMEC project and the colleagues in IST and partners of the project, Tratolixo, Selenis (today, Evertis) and PIEP (Department of Polymers of Minho University).

CONFLICT OF INTEREST

The author confirm that this article content has no conflict of interest.

REFERENCES

Billmeyer, F.W, Jr. (1984). *Textbook of Polymer Science*, Edited by John Wiley & Sons.

Burat, F., Güney, A., Kangal, M.O. (2009). Selective separation of virgin and post-consumer polymers (PET and PVC) by flotation method. *Waste Management, 29*, 1807–1813.

Carvalho, M.T., Rosa, L., Simões P., lvaro, Costa Á. (2007a). Semec Pilot Project: Mechanized Recovery of Granulated Plastics from Packaging. In: *Proceedings of the IFAC Workshop MMM 2007*, Quebec, Canadá.

Carvalho, M.T., Agante, E., Durão, F. (2007b). Recovery of PET from Packaging Plastics Mixtures by Wet Shaking Table. *Waste Management Journal, 27*, 1747-1754.

Carvalho, M.T., Ferreira, C., Portela, A., Santos, J.T. (2009). Application of fluidization to separate packaging waste plastics. *Waste Management, 29*(3), 1138-1143.

Carvalho, M.T., Durão, F., Ferreira, C. (2010). Separation of Packaging Plastics by Froth Flotation in a Continuous Pilot Plant. *Waste Management, 30*, 2209-2215.

Drelich, J., Payne, T., Kim, J.H., Miller, J.D. (1998). Selective froth flotation of PVC from PVC/PET mixtures for the plastics recycling industry. *Polymer Engineering Science, 38*(9), 1378-1386.

Fraunholcz, N. (2004). Separation of waste plastics by froth flotation—a review, part I, *Minerals Engineering, 17*(2), 261-268.

Fraunholcz, N., Kooijman, S.J.M., Dalmijn W.L. (2000). Flotation of plastics using combined treatments. In: *Proceedings of XXI International Mineral Processing Congress*, Rome, B12a-49–B12a-55.

Le Guern, C., Conil, P., Houot, R. (2000). Role of calcium ions in the mechanism of action of a lignosulfonate used to modify the wettability of plastics for their separation by flotation. *Minerals Engineering, 13*, 53–63.

Marques, G.A., Tenório J.A.S. (2000). Use of froth flotation to separate PVC/PET mixtures. *Waste Management, 20*, 265–269.

Pascoe, R.D., Hou, Y.Y. (1999). Investigation of the importance of particle shape and surface wettability on the separation of plastics in a Larcodems separator. *Minerals Engineering, 12*(4), 423-431.

Pongstabodee, S., Kunachitpimolb, N., Damronglerd, N. (2008). Combination of three-stage sink–float method and selective flotation technique for separation of mixed post-consumer plastic waste. *Waste Management, 28*, 475-483.

Santos, L., Paiva, M.C., Machado, A.V., Bernardo, C. (2007). Analysis of plastics from municipal solid waste by thermogravimetry and fourier transform infrared spectroscopy, In: *Proceedings of Materiais 2007*, Porto, Portugal.

Shen, R.J., Pugh R.J., Forssberg, E. (1999). A review of plastics waste recycling and the flotation of plastics. *Resources, Conservation and Recycling, 25*, 85–109.

Wills, B. (1998) *Mineral Processing Technology*, Elsevier, Amsterdam.

Send Orders of Reprints at reprints@benthamscience.org

CHAPTER 3

ADR: A Classifier for Fine Moist Materials

W. de Vries and P.C. Rem[*]

Department of Resources & Recycling, Delft University of Technology, The Netherlands

Abstract: MSWI-bottom-ash (IBA) and C&D-waste (CDW) are notoriously difficult to classify or separate at grain sizes below 12 mm. The problem is caused by the combined presence of (-1 mm) fines and moisture, which act together to form agglomerates and foul screens and other separation equipment. Results are presented of an experimental study into the separation performance of a new type of classifier, called Advanced Dry Recovery (ADR), which is designed to deal with fine moist materials. The study focused on IBA and two types of CDW: crushed concrete and sieve sand. For each of these three materials, the particle size distribution and moisture content of the fine and coarse ADR products were analyzed. It was found that more than 80% of the fines of the input materials is recovered into the fine product. For IBA, the reduction of fines in the coarse product resulted in more than double of the -8 mm non-ferrous metal recovery at the downstream eddy current separators. The coarse products from crushed concrete showed sufficiently low levels of cement fines, wood and foams to consistently satisfy the European norm for secondary aggregates and avoid problems with caking in storing this material. In a final experiment, about 23% of sieve sand was reclaimed as reusable aggregate.

Keywords: Bottom ash, Construction and demolition waste, Recycling.

INTRODUCTION

As a step towards a more sustainable society, technologies should be developed to recycle municipal solid waste Incinerator Bottom Ash (IBA) and Construction and Demolition Wastes (CDW) into high grade metal and mineral products instead of road foundation materials (Bertolini *et al.*, 2004, Gerven *et al.*, 2005a, Mulder *et al.*, 2007, Rao *et al.*, 2007).

Incinerator bottom ash has a high non-ferrous (NF) metal content (Chimenos *et al.*, 1999, Wiles and Shephard, 1999), which is presently recovered successfully only from the +12 mm fraction (BAT, 2004). Therefore it is interesting to recover

---

*Address correspondence to P.C. Rem: Department of Resources & Recycling, Delft University of Technology, Postbus 5048, 2600 GA, Delft, The Netherlands; Tel: +31(0)1527 83617; Fax: +31(0)1527 88162; Email: P.C.Rem@tudelft.nl

Vincenzo Gente and Floriana La Marca (Eds)

these metals also from the -12 mm fraction, both from an economic as well as from an environmental point of view (Birgisdottir *et al.*, 2007).

Aluminum is of interest because of its high scrap value and large carbon footprint. The recovery of heavy non-ferrous metals from the -12 mm fraction increases the overall recycling rate of these metals, resulting in less depletion of natural resources, and less metal content in the IBA, that may be a problem in the long term because of its possible leaching to the groundwater (BAT, 2004).

In the coming 20 years, a strong increase of the amount of construction and demolition wastes is expected in Europe because of the construction boom in the 1950's and the shortening lifespan of buildings. At the same time, the current main outlet for CDW, road foundations, is expected to shrink. The reason is that the net growth of infra-structure will diminish, resulting in a decreasing need for new road foundation material.

In principle, the surplus of waste can be used to advantage by recycling concrete from the CDW-stream into high-grade construction products, in particular to aggregates and cement. So far, however, even simple contaminants like wood and steel are not being removed from the -12 mm fraction, since this is difficult to achieve without complete drying or using an expensive wet process (Huang *et al.*, 2002).

To enable high-grade recycling, an efficient classification according to particle size is generally necessary prior to material separation. Yet, the typical fines and moisture contents of IBA and CDW make it difficult to screen the -12 mm fraction of these waste typologies.

Drying the materials to lower the moisture content consumes too much energy and wet methods produce sludge which has to be treated or land-filled, often at considerable costs. Aging and dry-season processing of IBA do improve its processability, but the recovery of metals remains unsatisfactory (Arickx *et al.*, 2006, Chimenos, 2000, Gerven *et al.*, 2005b).

The problem in treating the -12 mm fraction of IBA and CDW is primarily associated with the combined presence of -1 mm fines and moisture. This combination promotes agglomeration and makes the entire -12 mm fraction clumpy and difficult to separate. Reduction of the -1 mm content would make the remainder processable by screens, magnets, eddy current separators (ECS) or wind sifters. Once these conventional techniques become available, it is then possible to recover high grade metal and mineral products from IBA and CDW.

A new dry technology called Advanced Dry Recovery (ADR) separates moist materials with cut sizes down to 1 mm. An ADR-unit uses kinetic energy to break the bonds that are formed by moisture and fine particles (Berkhout and Rem, 2009; de Vries *et al.*, 2009). After breaking up the material, the fine particles are separated from the coarse particles. The resulting coarse product is then suitable for conventional upgrading processes. In order to investigate the effectiveness of ADR, two installations, of 40 t/h and 120 t/h respectively (Fig. **1**), were tested on IBA and two types of CDW: crushed concrete and sieve sand.

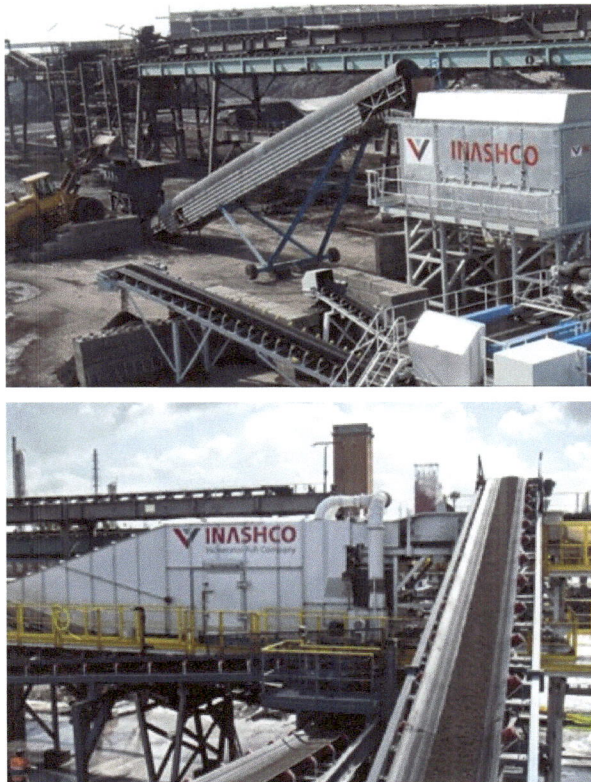

Figure 1: ADR installations at 40 ton/h (left) and 120 ton/h (right).

STATE OF THE ART

MSWI Bottom Ash

Currently, the standard technology for treating IBA is to remove all NF and ferrous metals above 12 mm using magnets and eddy current separators. The fine fraction is either land-filled or aged with the rest of the mineral product for 6 weeks to induce carbonation and reduce leaching. After acceptable leaching values are reached, the material can be used in road foundations. Depending on the local legislation, the material may need to be sealed, stored in a recoverable way and monitored for its entire lifespan (BAT, 2004).

Tables **1** and **2** present reference data for the processing of one ton of raw IBA by a Dutch state-of-the-art bottom ash processing facility operated at maximum aluminum recovery (and a relatively high slag content of the NF mix of 57%). Within the reference facility, three streams are produced. Table **1** shows the wet amount and moisture content of these three streams. The size distribution and aluminum content of the dry IBA-rest and NF metal concentrate are shown in Table **2**.

Table 1: Properties of main product streams

	Weight (kg)	Moisture (%)
IBA-rest	875	13.3
NF	41	6.1
Ferrous	84	-
Total	1,000	-

Table 2: PSD and aluminum contents/recovery

Size class	Size distribution (%)		Al grade (%)		Al recovery (kg)		Total recovery
(mm)	Rest	NF-Conc.	Rest	NF-Conc.	Rest	NF-Conc.	(%)
>20	7.8	44.3	0.1	27.6	0.08	4.69	98.4
6-20	16.3	29.5	0.5	49.7	0.57	5.62	90.8
2-6	30.0	11.8	1.3	39.8	3.06	1.80	37.1
<2	45.9	14.4	0.7	0.6	2.61	0.03	1.27
Total	100	100	0.8	31.6	6.32	12.14	65.8

The data show an excellent aluminum recovery for the +6 mm class. However, the recovery for the 2-6 mm is significantly lower. One ton of wet IBA contained about 18.5 kg of aluminum. Almost 3.1 kg remains in the 2-6 mm fraction of the IBA-rest and is not recovered. Separation of this fine aluminum using eddy current separators is theoretically possible; however properties of the fine material cause it to be uneconomic.

C&D Waste

The current primary use of construction and demolition waste is in road foundations. Only a minor part (<5%) is processed in order to be used as aggregates in concrete production. For crushed concrete to be used as aggregate it is necessary to remove fines as well as contaminants such as foams, wood and steel. The best available technique to process CDW into aggregates is washing or complete drying combined with wind-sifting. Both methods are expensive and energy consuming (RILEM TC 165-SRM, 2000). In an earlier study concerning the high grade recycling possibilities of CDW, it was tried to separate the different materials in CDW using thermal treatment, followed by material classification *via* sieving and wind-sifting. Because the thermal treatment was carried out on almost the entire CDW stream, it was found to be uneconomic (Mulder *et al.*, 2007).

RESEARCH METHODOLOGY

The performance of the ADR was tested in four experiments, using different input materials. One of the tests used MSWI Bottom Ash (IBA) with a size range of 0-8 mm. The others were carried out on the following three typologies of Construction and Demolition Waste (CDW):

- A heavily polluted crushed concrete;

- A crushed concrete with a low pollution grade; and

- Sieve sand.

For the four materials the moisture content and particle size distribution of the input was determined and compared to the same data for the coarse and fine product produced by the ADR. For IBA, also the recovery of non ferrous metals after the ADR was determined and for the CDW-materials the amount of floating contamination (*i.e.* wood) in the +4 mm size was measured according to the method of the Dutch NEN-EN 933-11. An overview of the composition of the input materials is presented in Table **3**. The Particle Size Distributions (PSD) of the four materials are shown in Fig. **2**.

Table 3: Main properties of processed IBA and CDW materials

IBA and CDW Materials		Moisture	0-1 mm	Non Ferrous	Floating (+4 mm)
		(kg/kg wet)	(kg/kg)	(kg/kg dry)	(cm^3/kg)
MSWI Bottom Ash	(IBA)	17.0%	28.4%	1.2%	-
Polluted Crushed Concrete	(PCC)	11.7%	47.1%	-	12.3
Clean Crushed Concrete	(CCC)	8.4%	30.4%	-	1.38
Sieve Sand	(SS)	20.5%	53.0%	-	215.5

Since the ADR is able to classify raw bottom ash, there is no need to age the material before processing. Aging the ash creates heat which evaporates moisture, making the bottom ash better processable. This heat is most likely generated by the oxidation of aluminum, reducing the recoverable mass of non-ferrous metals. In order to assess the effect of aging on the aluminum content and metal recovery, two samples of IBA were analyzed. One of these samples had not been aged; the other sample had aged for 10 weeks.

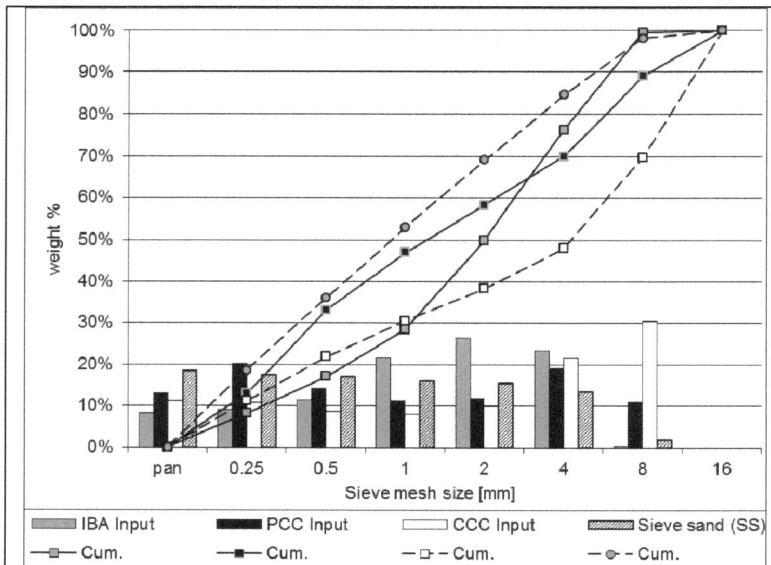

Figure 2: Size distribution of IBA and CDW-materials.

EXPERIMENTAL RESULTS

MSWI Bottom Ash

The 0-8 mm MSWI bottom ash was classified at 1 mm into a fine product consisting of 38% wt of the input and a coarse product of 62% wt of the input. Fig. **3** shows the particle size distribution of the input, fine and coarse product. It is also shown the control series, which is the weighed average of the fine and coarse product. This control is slightly finer than the input material, because some of the particles break during processing.

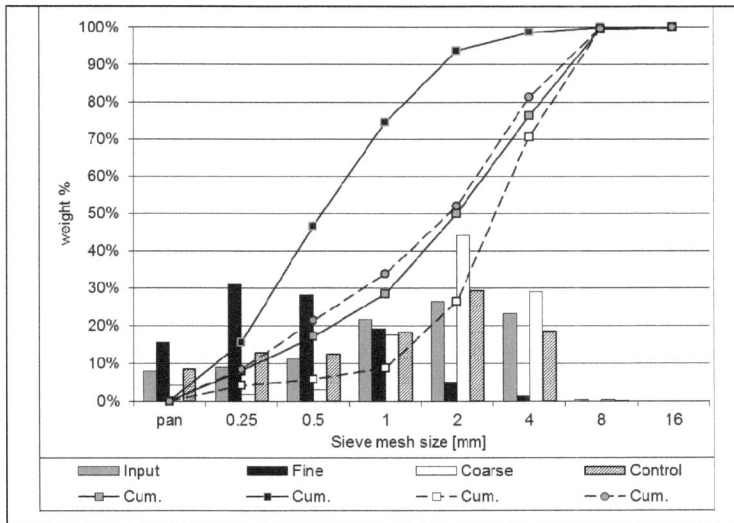

Figure 3: PSD of 0-8mm MSWI bottom ash processed with the ADR.

The data show that the two products have clearly distinct particle sizes. Since moisture is associated with fines, the moisture content of the coarse fraction decreased from 17% to 13%. These two effects (reduction of fines content and moisture content) result in a material that can be well processed using Eddy Current Separators (ECS).

Images of the input, fine and coarse product of IBA are presented in Fig. **4**. The input mix of fine and coarse particles shows a surface coating of very fine particles on all grains because of the moisture that is present. After processing, the adhering fines are removed from the coarse product, resulting in a loose and processable mixture.

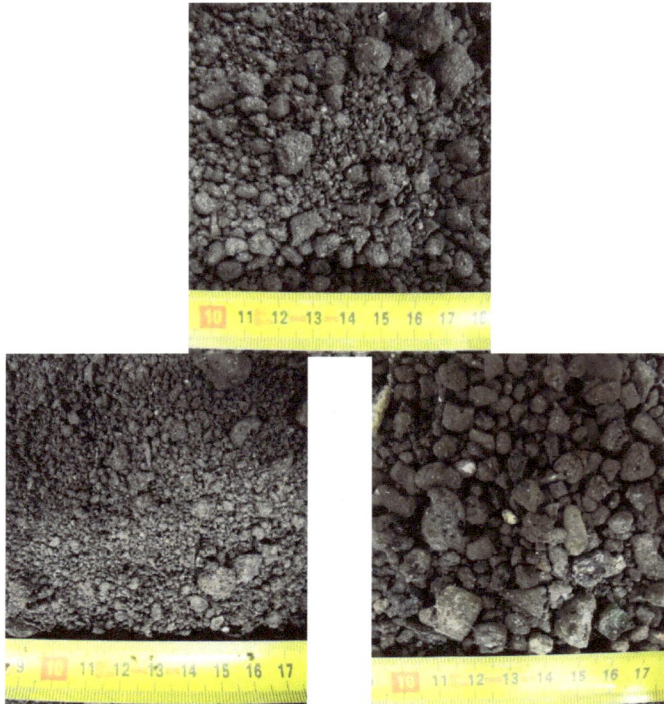

Figure 4: Bottom ash input, fine and coarse product.

The coarse material produced by the ADR was further processed using industrial eddy current separators to concentrate the non ferrous metals.

Results showed a high recovery of 89% of NF metals in the 1-8 mm product (twice the recovery of the state of art reference case). It should be noted that the input material used in this experiment (0-8 mm IBA) had a lower grade of NF than the state of the art reference case, because it had been aged for 6 weeks.

The amount of fines in the mineral reject of the ECS is strongly reduced, making it more suitable for road foundation. Furthermore, since more NF metal is recovered by the ECS, the metal content in the mineral product, and with it the ultimate leaching potential, is reduced, improving its properties when applying the material as a foundation layer. The fine fraction produced in the ADR (-1 mm), can be an interesting material in the production of cement (Pan *et al.*, 2008).

Currently bottom ash is aged to reduce the moisture content and enable better processing. It is known from general practice that during this aging the temperature increases from 40°C to around 80°C, which is suspected to be caused by the oxidation of aluminum. To confirm this, the aluminum content of fresh and aged bottom ash was determined.

Table **4** presents the properties of fresh and aged IBA.

Table 4: Moisture content and aluminum grade found in fresh and aged IBA

	Fresh	Aged
Moisture content	15.4%	13.0%
Total metallic aluminum grade	2.7%	1.8%
Recoverable aluminum	2.3%	1.0%
Unrecoverable aluminum	0.4%	0.8%

During aging the moisture content reduces by 2.4% and the total metallic aluminum content by one third. The energy of evaporation needed for the observed reduction of the moisture content is in accordance with the energy released by the oxidation of aluminum. This supports the hypothesis that aluminum oxidation is the main cause for drying the ash during aging.

During aging the oxidized aluminum forms a layer around the aluminum particles. This layer causes the total aluminum particle to be less recoverable and reduces the recoverable amount of aluminum by 56%.

C&D Waste

Three different types of CDW were processed using the same industrial installation of the ADR. Figs. **5**, **6** and **7** show the particle size distributions for heavily polluted crushed concrete, relatively clean crushed concrete and sieve sand, respectively.

For all three materials the distinction between the coarse and fine product is very clear, the D_{10} of the coarse product increased by almost a factor eight with respect to the input. This reduction in fine classes is accompanied by a reduction in moisture content of about one quarter, similarly as in the experiment with bottom ash.

However, since the absorbed moisture is of no concern in processing the material (absorbed moisture plays no part in particle-particle interaction), the effect on the

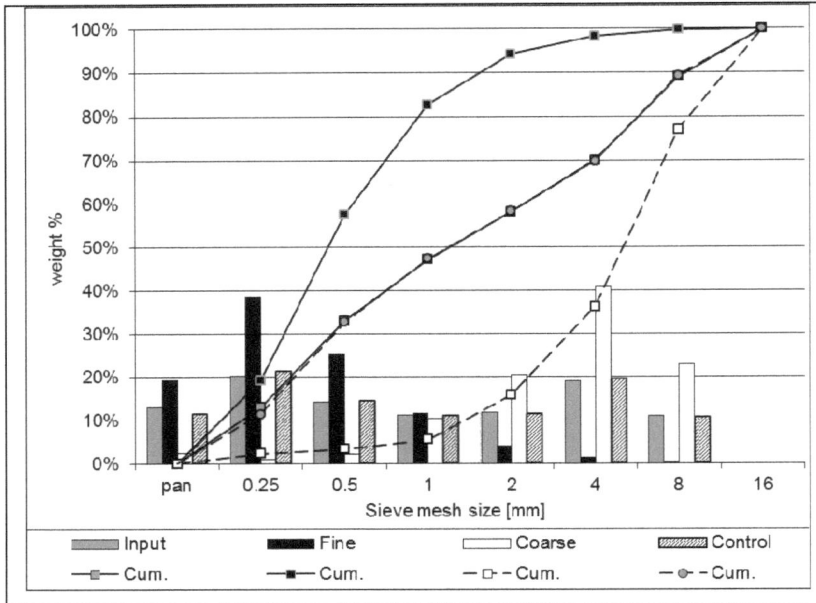

Figure 5: Heavily polluted crushed concrete.

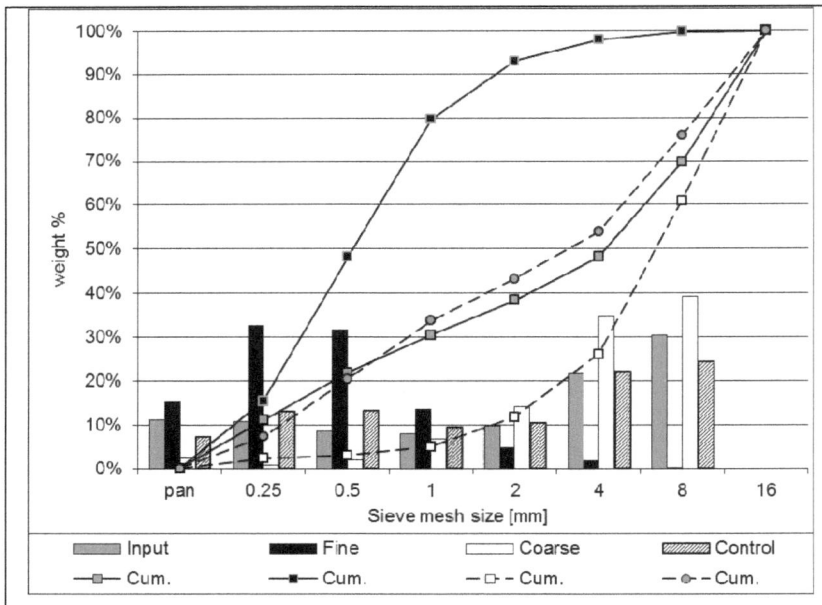

Figure 6: Relatively clean crushed concrete.

free moisture is actually larger. Ordinary (non-crushed) concrete has about 5% of absorbed moisture because of the cement hydration process. At a typical crushed concrete bulk moisture level of 15% there will be 10% adsorbed water, which is reduced to 6% in the coarse product. For crushed concrete further processing may be needed to make it suitable for use as aggregates in recycled concrete. For this application there are a number of criteria, of which two are most critical. The workability of the bulk material (*i.e.* the ability to transport and store the material) sets boundaries to the amount of fines, and the concentrations of contaminations, mainly the floating contaminations, limit the application of the material as an aggregate in concrete.

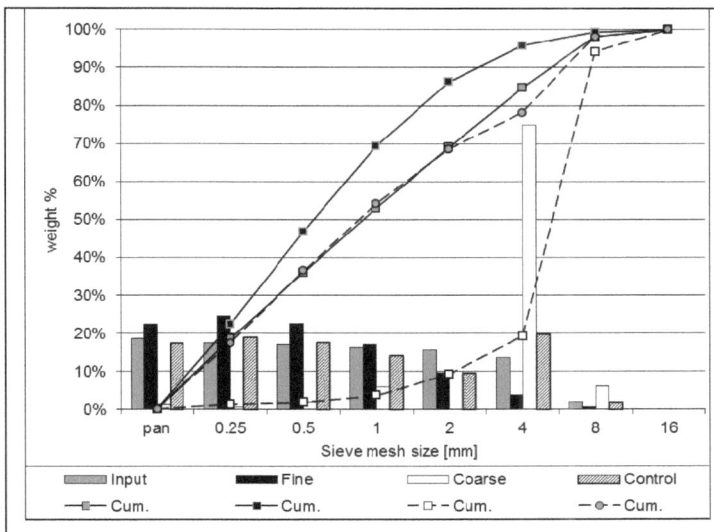

Figure 7: PSD of processed sieve sand.

Most of the light contaminants are also concentrated into the fine product. This effect was quantified following the method described in the Dutch NEN-EN933-11. The procedure entails the separation of light weight material larger than 4 mm by means of floatation. The volume of floating material is then measured using a graduated cylinder. The results of this analysis are shown in Table **5**.

Table 5: Amount of floating material in +4mm fraction

	PCC [cm³/kg]	CCC [cm³/kg]	SS [cm³/kg]
Input	12.3	1.38	215.5
Fine	233.1	63.62	684.5
Coarse	4.0	0.04	71.3

The contamination levels of the coarse products of both types of crushed concrete are well within prescribed categories for application of granulate as concrete aggregates. However, the value for the coarse product of the heavily contaminated crushed concrete is too high to satisfy the strictest categories. Further processing may therefore be desirable. One route is wind sifting, to remove all light particles in the size range of 0-16 mm. Because of their lower specific weight, light pollutants will be associated by 1-4 mm mineral material. This 1-4 mm mineral fraction (which then contains all contaminations) can be processed in a wet process (*e.g.* a screw classifier) to remove all floating material. Using this technique only 10 to 15% of the total 0-16 mm crushed concrete needs to be cleaned in a wet process. Since the amount of fines has already been significantly reduced, problematic amounts of sludge are avoided.

Clean coarse products of CDW can be used as aggregate in concrete. In this way it will be possible to close the cycle of concrete aggregates, reducing the amount of unused waste material and depletion of natural resources. The fine material produced by the ADR is a promising material for the cement industry.

Figs. **8** and **9** show images of crushed concrete and sieve sand that have been processed to form a fine and coarse product. Since sieve sand is a very complex waste stream, it is preferred to reduce it as much as possible. Classifying sieve sand is therefore an interesting application of the ADR technology. Using the ADR, 23% of the input mass was classified into a coarse product, which can be used as an ordinary granular mixture in various applications.

Table **6** shows the mass percentage, moisture content, grades and recoveries of the input, coarse and fine product for each of the four experiments.

Fig. **10** shows the recovery of every size fraction in the coarse product for all four series. The different series in this graph show good and comparable classification efficiencies for all four series (same slope around splitting point); only the cut point varies. In these experimental series, the settings of the ADR were not adjusted, so the change in cut point is a result of the properties of the processed material.

Figure 8: Concrete granulate input, fine and coarse product.

Figure 9: Sieve sand input, fine and coarse product.

Table 6: Summary of results and recovery rates

		IBA [%]	PCC [%]	CCC [%]	SS [%]
Mass distribution	Fine	38.0	54.1	38.5	77.0
	Coarse	62.0	45.9	61.5	23.0
Moisture content	Input	17.0	11.7	8.4	20.5
	Fine	22.3	13.7	11.2	21.6
	Coarse	13.3	8.4	6.7	14.6
Recovery of 0-1mm	Input	28.4	47.1	30.4	53.0
	Fine	74.6	82.7	79.7	69.3
	Coarse	8.8	5.5	5.0	3.6
Grade of 0-1mm	Fine	83.9	94.6	90.9	98.5
	Coarse	16.1	5.4	9.1	1.5
Recovery of >1mm	Fine	14.6	17.7	11.8	51.6
	Coarse	85.4	82.3	88.2	48.4

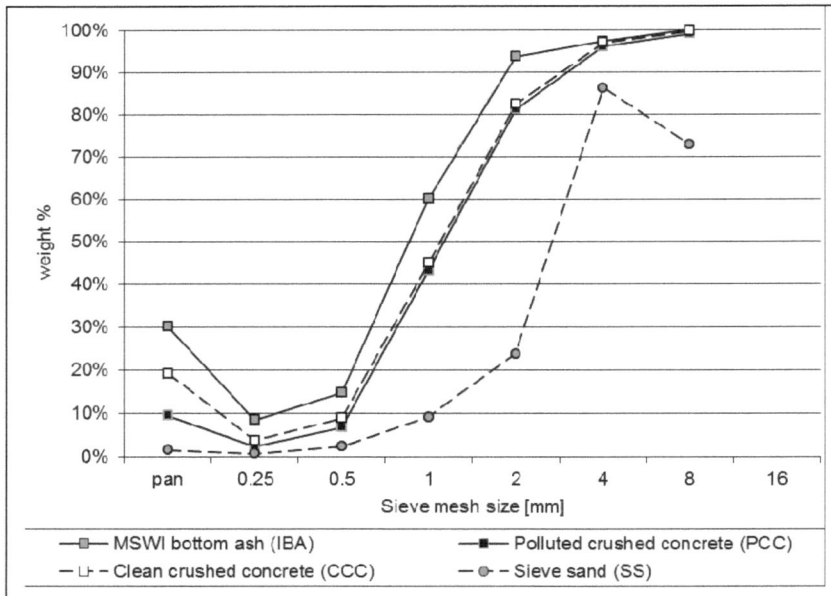

Figure 10: Recovery of every size fraction in the coarse product.

CONCLUSIONS

Problems in the classification and separation of materials from the 0-12 mm fraction of MSWI-Bottom-Ash (IBA) and Construction & Demolition-Waste

(CDW) are caused by (-1 mm) fines and high moisture levels associated with these fines. An experimental study has been done into a new type of classification technology, called the Advanced Dry Recovery (ADR) that removes most of the -1 mm fines from wastes. The experiments showed a reduction of 84% in the -1 mm content for incinerator bottom ashes. For CDW a reduction of more than 90% was found.

After the reduction of fines and moisture, valuable materials or problematic contaminants can be separated from the wastes by screens and eddy current separators. In particular, the recovery of non-ferrous in the -8 mm fraction of IBA doubled, compared to a state-of-the-art reference case. Moreover, if the fines are removed from IBA, aging for several weeks can be avoided, which significantly improves the recovery of aluminum. Classification of the CDW mineral fraction at 1 mm or 2 mm also significantly reduces floating contamination. In particular, a course granular product can be made from sieve sand, a highly complex kind of CDW.

ACKNOWLEDGEMENT

Declared none.

CONFLICT OF INTEREST

The author confirms that this chapter content has no conflict of interest.

REFERENCES

Arickx, S., van Gerven, T., Vandecasteele, C. (2006). Accelerated carbonation for treatment of MSWI bottom ash. *Journal of Hazardous Materials*, *137*(1), 235-243.

Berkhout, S.P.M., Rem, P.C. (2009). Dutch Patent NL2001431.

Bertolini, L., Carsana, M., Cassago, D., Curzio, A.Q., Collepardi, M. (2004). MSWI ashes as mineral additions in concrete. *Cement and Concrete Research*, *34*(10), 1899-1906.

Best Available Technology, BAT, (2004) *Treatment techniques for solid residues*. Document PJ/EIPPCB/WI, Chapter 4.6, European Union, 388-390.

Birgisdottir, H., Bhander, G., Hauschild, M.Z., Christensen, T.H. (2007). Life cycle assessment of disposal of residues from municipal solid waste incineration: Recycling of bottom ash in road construction or landfilling in Denmark evaluated in the ROAD-RES model. *Waste Management*, 27(8), S75-S84.

Chimenos, J.M., Segarra, M., Fernandez, M.A., Espiell, F. (1999). Characterization of the bottom ash in municipal solid waste incinerator. *Journal of Hazardous Materials*, 64(3), 211-222.

Chimenos, J.M., Fernandez, A.I., Nadal, R., Espiell, F. (2000). Short-term natural weathering of MSWI bottom ash. *Journal of Hazardous Materials*, 79(3), 287-299.

van Gerven, T., Geysen, D., Stoffels, L., Jaspers, M., Wauters, G., Vandecasteele, C. (2005a). Management of incinerator residues in Flanders, Belgium and in neighbouring countries. A comparison. *Waste Management*, 25(1), 75-87.

van Gerven, T., van Keer, E., Arickx, S., Jaspers, M., Wauters, G., Vandecasteele, C. (2005b). Carbonation of MSWI-bottom ash to decrease heavy metal leaching, in view of recycling. *Waste Management*, 25(3), 291-300.

Huang, W.L., Lin, D.H., Chang, N.B., Lin, K.S. (2002). Recycling of construction and demolition waste *via* a mechanical sorting process. Resources. *Conservation and Recycling*, 37(1), 23-37.

Mulder, E., de Jong, T.P.R., Feenstra, L. (2007) Closed cycle construction: An integrated process for the separation and reuse of C&D waste. *Waste Management*, 27, 1408-1415.

Pan, J.R., Huang, C., Kuo, J.J., Lin, S.H. (2008). Recycling MSWI bottom and fly ash as raw materials for Portland cement. *Waste Management*, 28(7), 1113-1118.

Rao, A., Jha, K.N., Misra, S. (2007). Use of aggregates from recycled construction and demolition waste in concrete. *Resources, Conservation and Recycling*, 50(1), 71-81.

RILEM T.C 165-SRM. (2000). *Report rep022: Sustainable Raw Materials – Construction and Demolition Waste – State-of-the-Art Report of RILEM*, Editors: Hendriks, CF, Pietersen, HS.

de Vries, W., Rem, P.C., Berkhout, S.P.M. (2009). ADR: a new method for dry classification. *Proceedings of the ISWA International Conference, Lisbon*, 12-15 October 2009, 103-113.

Wiles, C., Shepherd, P. (1999). *Beneficial use and recycling of municipal waste combustion residues – A comprehensive resource document*. NREL, report BK-570-25841.

Send Orders of Reprints at reprints@benthamscience.org

Separating Pro-Environment, 2012, 59-76

CHAPTER 4

Sensor Based Sorting in Waste Processing

J. Julius and Th. Pretz[*]

Department of Processing and Recycling, RWTH Aachen University, Aachen, Germany

Abstract: Sensor based sorting consists of the separation of single particles on the basis of identifiable material attributes that can be measured by suitable detectors. This technique has revolutionized the design of mechanical treatment processes especially in the field of dry waste separation systems. The complexity and the number of sensors are the main determinative factors influencing the separation results. Today, newly developed devices are often equipped with a combination of different sensors in order to recognize multiple material properties in a single step, whereas, the sensor data are evaluated in real-time by a computer system. As a consequence, these devices guarantee excellent separation efficiencies in comparison with those having single sensors. Additionally, the advancing development of new sensor systems makes new fields of application in waste treatment accessible.

Keywords: Sensor sorting, Waste processing, Resource recovery, Selective separation, Recycling.

INTRODUCTION

The development of techniques for waste recycling exhibits a history which started approximately forty years ago. Mechanical processing methods to be adopted for this purpose consist of the following main unit operations: comminution, size classification and separation.

Comminution is predominantly used for the reduction of the upper grain size and for liberation of composite materials in the feed stream.

Size classification or screening aims at obtaining different material flows in respect to particle dimensions. In this way it is possible to reduce the mass rate of flows to be treated within the processes, to provide feed materials in suitable grain size ranges

*Address correspondence to Th. Pretz: Department for Processing and Recycling, RWTH Aachen University, Wüllnerstraße 2, 52062 Aachen, Germany; Tel.: +49(0)2418095700; Fax: +49(0)2418092624; Email: pretz@ifa.rwth-aachen.de

for subsequent process stages and to obtain final products in defined grain size ranges.

Finally, separation procedures are utilized for the segregation of feed mixtures into different material streams according to suitable material attributes. These attributes can be *e.g.* density, shape, magnetizability or electrical conductivity. Initially, separation methods were almost exclusively adopted from the ones that constituted common practice in conventional mineral processing. However, the specific properties of complex waste mixtures, like low bulk densities, poor aptitude of bulk commodities and mostly wide variable particle shapes required the development of specific processing systems. Thus, newly designed sorting devices, like eddy current separators, were able to capture the recycling equipment market.

An important improvement in separation technique succeeded with the implementation of sensor based sorting. This method can be characterized by the possibility of separating single grains on the basis of externally identifiable material attributes that are measured by suitable detectors. This processing method benefited from the rapid development of sensors and digital data handling. In the first phase of the application, approximately at the beginning of the 1990's, these devices were adopted to process separately collected recyclable materials, like waste glass and light packaging waste. For instance, sensor based sorting allowed recognizing different colours by means of transmission measurement or different kinds of plastics with near infrared (NIR) spectroscopy. Moreover, as an additional advantage sensor based sorters allowed the substitution of the widely-used hand picking with increasingly powerful automatic systems. Thus, the identification of material attributes within waste mixtures was considerably improved in comparison with the deficient human vision. For a widespread launching of sensor based sorting several years occurred mainly because it was necessary to find the solutions to adapt the technical performances of sensor based sorting devices to the quality demands of the recycled materials.

Today, several sensor based sorters are available on the market. These devices open up, in many cases, completely new separating criteria thus providing a more effective waste separation. For these purposes specialized sensors are successfully used for material characterization, *e.g.* with X-ray transmission measurement or digital image processing. Sensor based sorting revolutionized the design of separation processes for the treatment of waste materials especially in the field of dry handling systems. Hence, the development timeframe of approximately twenty years resulted in a considerable innovation in the field of separation devices with contactless detection. This permits to accomplish the aim of obtaining quality assured final products from waste.

FUNDAMENTALS OF SENSOR BASED SORTING

Separation technology can be divided into direct and indirect systems. Direct systems make use of interactions between specific properties of single particles and specific force fields applied by the separating devices. A typical example of direct separation is magnetic separation which allows recovering selectively iron and steel thanks to different susceptibilities of material components in the waste feed stream.

In contrast sensor based sorting is an indirect separation. A typical example is the application of this method for identification of material characterised by different colours in the waste feed stream by means of a specific detector. The separation occurs subsequently by deflection from the waste flow of selected particles, preferably with air blasts from blowing nozzles.

Sensor based sorting, also known as automatic picking, exhibits a separation method which can be characterized by single particle recognition by means of externally and contactless identifiable attributes which are measured with specialized detectors. These material attributes may be *inter alia* colour, shape, brilliance, molecular composition, electrical conductivity and density of matters (Killmann and Pretz, 2006).

In general, sorting devices which operate according to this principle consist of the following components:

- Transportation devices for feeding and singling of materials

- Emitter systems for transmitting electromagnetic radiation towards the particles to be separated,

- Sensor systems for recognition of specific material properties,

- Software-controlled data processing for interpretation of the signals coming from the sensors and

- Mechanical discharge units, in most cases equipped with a pneumatic manifold with electromagnetically actuated air jets for blowing out identified particles.

Newly developed sensor sorters are often outfitted with a combination of different sensors for a simultaneous recognition of several material attributes. The data collected by these sensors are processed and analysed by means of computers in real-time. Thus, sorting devices equipped with multiple sensor systems guarantee better separation efficiencies than sorters with single detectors. This is especially true for the separation of complexly composed waste mixtures with a high variety of qualitative characteristics. Another advantage of modern sensor based sorters is the learning ability resulting from software-controlled data processing. This characteristic permits an increased adaptability especially with respect to modified process flow sheets, changing of input material compositions or of requirements concerning the product qualities for the recovered materials. Moreover, in comparison with conventional separation systems, sensor based sorters provide a decoupling of the actual separation so that the danger of carrying over and thus cross-contamination of products can be minimized.

The main requirement to attain good separation results with sensor based sorters is a convenient conditioning of the feed material. As a rule, this includes a pre-classification with screens in order to provide particle size ranges as narrow as possible. Additionally, screening is also necessary to separate fine grained and coarser particles as these are often too small or too big for the adopted sensor sorting system. Preceding separation stages with conventional separating techniques can be frequently advantageous in order to increase the grade of valuable materials in the feed flow for the sensor sorters and to remove troublesome components like plastic foils and other light materials *e.g.* by air classification.

The separation efficiency of sensor sorters depends to a great extent on the quality of singling on the transportation device in the region of the sensor and discharge system. Thus, it is necessary to achieve a single layer whereas particles should not touch or even overlap each other in order to avoid misplaced materials in the final products. The most common method to accomplish this aim is to apply a cascade of belt conveyors and/or inclined sliding chutes with rising transport velocities.

Moreover, the particles to be separated must have a stable position for instance on a belt conveyor as relative movements of single grains make a time- and position-dependent determination of the actual blowing nozzles to be activated impossible. Positive as well as negative separation can be conducted with automatic sensor sorting. Positive separation means that recyclables are enriched in the accepted valuable product while negative separation describes the well-directed removal of bothersome constituents.

Most of the sensor based sorting devices utilize a positive separation which is provided by blasts of compressed air from nozzles whose valves are activated by software-controlled systems. Only very few sorter devices adopt mechanical discharge mechanisms like electromagnetic driven pushers (Fears, 2008). These systems have the advantage that the cost extensive generation of compressed air can be omitted. However, their reaction speed is lower than the pneumatic discharge devices. Therefore, due to the lower transport velocity, the achievable throughput also drops.

Machinery Equipment for Sensor Based Sorting

The automatic machines used for sensor based sorting may be divided into the basic construction kinds of sliding chute and belt conveyor types. The sliding chute machines were already developed at the beginning of the 1980's. At first, they were almost exclusively employed for the separation of glass cullet according to different colours or transparencies. Meanwhile, these types of machines have reached a high capability and are particularly applied for sorting of fine grained materials as well as free-flowing bulk materials. Using the example of colour separation, Fig. **1** shows the typical design and the functioning principle of chute type sorters. In these kinds of systems, a first a vibrating conveyor (1) provides the basis to achieve an equal distribution of the grains of the feed material over the whole working width and to obtain a pre-singling. Following the grains are accelerated and further disseminated on the inclined chute. At the end of the chute the inspection of different colours is carried out with a line scan camera (2) and an associated illumination system, for instance consisting of two bars equipped with white light-emitting diodes (LED's).

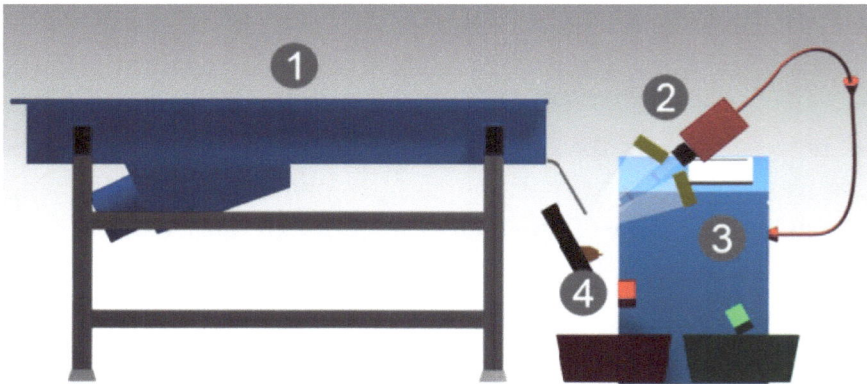

Figure 1: Principle of sensor based sliding chute type sorter (Pretz and Julius, 2008).

The line scanning sensor creates an endless image of the material stream which is classified by a computer system (3) according to the colour, position, shape and size of recognized particles. Thus, the computer system with a delay of few milliseconds activates the air pressure valves (4) in correct position allowing to blow out positively detected particles and discharge them over a product splitter. The programmable software of the computer allows adjusting in a wide spectrum the classification in different colours. Marginal differences in colour as well as in brightness which cannot be identified by human eyes are sufficient for precise recognition.

As most of the components contained in waste mixtures do not meet the property "free-flowing bulk solids" the so-called belt conveyor types for sensor sorting systems have been developed. These separating devices are also applied for coarse and irregular shaped materials, they are designed according to the scheme displayed in Fig. **2** and can be equipped with different kinds of sensors. The following description corresponds to a sorter with near infrared detection (NIR sensor). In most applications, the material is fed onto a vibrating conveyor (not shown in Fig. **2**) fulfilling the tasks of particle spreading and dissemination. The vibrating conveyor is followed by an inclined chute (1) on which the particles are sliding down and are accelerated till they reach approximately the velocity of the subsequent belt conveyor. This belt conveyor runs with a speed of in the range of 2.5-3.0 m/s. The continuously increasing conveying velocities produce a good singling of particles on the belt conveyor.

Above the belt conveyor, two rows of infrared emitters equipped with halogen lamps as well as a NIR sensor (2) are installed. The NIR sensor scans the complete width of the belt and transmits a spectrum of the analysed materials to the connected computer (3). The acquired spectra are compared in real-time with characteristic spectra stored in a data base. Additionally, also position and size of the particles can be determined. In case of a positive identification of single particles one or more air nozzles (4) are activated by the electronic system in order to achieve a selective discharge over a splitter plate.

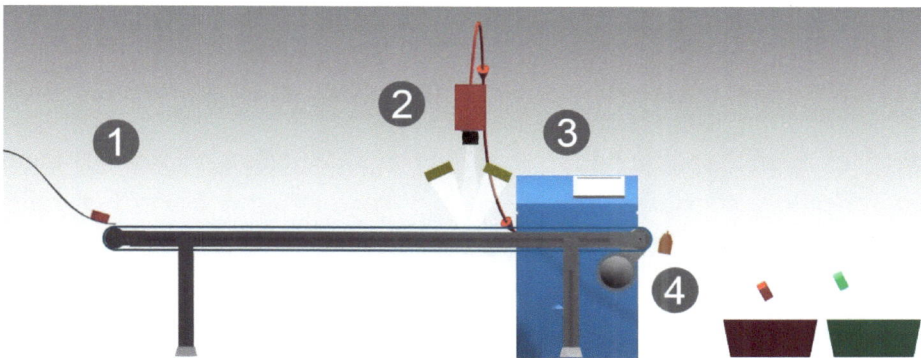

Figure 2: Principle of sensor based belt type sorter (Pretz and Julius, 2008).

The air nozzles are arranged in a distance of 10 to 30 mm in-line and parallel to the head pulley of the belt conveyor. The air blasts are triggered by the computer system which activates electromagnetic valves. Dependent on the grain size and shape of recognized particles several air valves can be activated.

The accuracy of recognition can be improved by the additional use of a second sensor. This may be for instance a colour line-scan camera which permits a more precise identification of grain sizes and shapes as well as a better determination of positions of single particles on the belt. Furthermore, the additional colour information can be utilized *e.g.* to separate plastics exhibiting a desired dye.

Finally, sensor based sorting devices are available on the market which include two rows of air nozzles for the discharge of two streams with different material attributes. Thus, it is possible with an adequate position of also two splitter plates to achieve a separation of two products with different attributes in one passage.

Sensor Systems and Spectral Ranges

The sensor systems of sensor based sorting devices typically consist of at least one emitter and one detector. The complexity of the detector is the decisive parameter for possible fields of application. The emitter is responsible for transmitting electromagnetic radiation towards the particles that have to be recognised. These particles affect the radiation selectively in respect to the spectrum, the time-dependent distribution, the amplitude and the direction of propagation. The characteristically modified radiation reaches the sensor system and is measured. The performance data are digitally recorded, further analysed with a computer and reduced to material attributes so that the software-controlled system can recognise the particles and eventually take action to separate them.

Regarding the spatial configuration of emitter and detector it can be distinguished between reflection, transmission and excited emission processes. For reflection measuring, emitters and detectors are arranged at the same side of the material stream. This procedure is applied, for instance, in the visible spectral range for colour sorting. If the particle flow is between emitter and detector, this is referred to as transmission gauging. For example, with this kind of system it is possible to analyse the different absorption of X-rays caused by solids. Finally, there are methods which use radiation of high energy densities. These kinds of systems adopting a stimulated emission allow the separation for instance of materials according to the characteristic of selective fluorescence (Killmann and Pretz, 2007).

The electromagnetic radiation transmitted by the emitter can be characterized, *inter alia*, by its spectral composition. The shorter the wavelength, the higher is the energy of the radiation. Sensor based sorting utilizes radiation in a wide range of the spectrum. Fig. **3** illustrates the different kinds of adopted radiation which extend from acoustic waves to gamma radiation. Most frequently, the visible (VIS) and near infrared light (NIR) as well as X-ray frequency ranges are applied. In the visible sector illuminants like fluorescent tubes, halogen bulbs and light emitting diodes (LED's) are used as emitters. The "white" light from these sources describes a mixture of colour gradations which are differently reflected and absorbed by the object surfaces. The shades of colour then are detected by charge-coupled devices (CCD) or complementary metal oxide semiconductors (CMOS) as sensors.

Gamma Rays (Wave Lengths < 1 pm)
Hard and Soft X-Rays (0,1 - 10 nm)
Ultraviolet Radiation (10 - 400 nm)
Visible Light (400 - 700 nm)
Near Infrared Light (700 - 1000 nm)
Infrared Radiation (1000 nm - 1 mm)
Microwave Radiation (1 nm - 1 m)
Radio Frequency Radiation (1 m - 1 km)
Acoustic Radiation (several kilometres)

Figure 3: Spectral ranges for emitters in sensor based sorting.

In the near infrared range, halogen lamps are most frequently employed as emitters. In fact, besides visual light, these sources exhibit a great amount of infrared radiation. Generally, the reflected light from the waste components is scanned with a rotating mirror system and identified with a sensor. In this way, the distribution of intensities of characteristic and well-known wave lengths is a direct attribute that can be used for the recognition of the different materials.

Sensor based sorters, equipped with inductive metal detectors, operate most frequently in the long-wave range of radio frequencies. The detectors consist of a coil system installed underneath a conveying device. These systems are made up of concentric coils each with a field coil and a receive coil. The operating principle is that the alternating magnetic field generated by the field coil is influenced by metallic particles. The receive coil detects the resulting field and compares phasing as well as amplitude with the same parameters of the generating field. Thus, electrically conductive materials like metals can be distinguished from non-conductive materials, even if only small portions of metals are contained, *e.g.* as it is with electrical components or insulated wires.

APPLICATIONS OF SENSOR BASED SORTERS

Actual developments in the field of sensor based sorting proved that a growing amount of processing functions can be covered with this technique. This is based on the continuous development of new detectors as well as on the implementation of multi-sensor systems.

Table **1** gives a non-exhaustive overview of the used separating criteria, sensor technologies and typical examples for applications. In order to prove the performance of modern sensor based devices and to demonstrate new developments, some applications are depicted in the following sections.

Table 1: Sensor based sorters and typical applications

Separating Attributes	Sensor Technology	Examples for Separation
Colour, brightness	Colour Cameras, VIS spectrometers	Waste glass, PET flakes, plastic granulates, copper and brass from non-ferrous metals, printed circuit boards from WEEE
Transparency	Colour Cameras	Opaque components from cullet
Colour, brilliancy	Colour Cameras	Magazines from waste paper
Molecular Composition on surface	NIR- spectrometers	Different kinds of plastics, beverage cartons, paper, cardboard, wood, textiles, PVC from refuse derived fuel
Electrical conductivity	Inductive detectors	Metals from diverse waste mixtures, stainless steel from metal mixtures
Density	X-ray transmission sensor	Aluminium and magnesium from metal mixtures, WEEE from waste mixtures, inert materials from waste wood, organics from building rubble
Chemical composition	Laser induced breakdown spectroscopy (LIBS)	Different magnesium, aluminium and steel alloys from metal mixtures

Separation in the Visible Spectral Range

The separation of waste components according to different colours is the oldest application of sensor based sorting devices and it was primarily adopted for the separation of waste glass as well as for the removal of opaque non-glass ingredients like ceramic, stones and porcelain from cullet. For these tasks sliding chute machines are mainly applied adopting one or two line scan cameras for the inspection of particles from one or two sides (Zeiger, 2005). In case of the selective separation of opaque non-glass components, transmission measurement is employed while an analysis *via* reflection is possible if white, green and brown cullets are to be separated.

Further fields of application for sensor based colour sorters are found in processing of different kinds of plastics. A typical example is the treatment of PET flakes which requires a high grade separation especially for the utilization in the fibre manufacturing industry. The content of colour impurities must not exceed more than 100 ppm (Zeiger, 2006). Furthermore, the content of other foreign materials has to be in the single-digit ppm-area. The adoption of high resolution CCD line scan cameras guarantees that particles greater than 2 mm as well as dark inclusions greater than 0.5 mm, *e.g.* in plastic granulates, can be definitely identified and the specifications concerning product purities can be met. Typical sorting machines for colour sorting of plastic flakes are mainly among the sliding chute type separators. Fig. **4** gives an impression of such a machine which has a working width of approximately 900 mm and a capacity of 1 - 3 t/h, mainly dependent on material composition and particle size distribution (Mogensen, 2004).

Figure 4: Mogensen MikroSort® for processing of plastic flakes (PET, PE, PP, PVC) (Dehler, 2008).

Separation in the Near Infrared Spectral Range

Sorting devices which utilize the NIR-range were used primarily for the separation of materials from light packaging waste. Today, a broad span of applications is given, *inter alia,* for processing of pre-treated household and commercial waste in the particle size range of 25 - 400 mm. In principle, with NIR sensors the following waste components can be identified and be selectively separated: beverage cartons, paper, cardboard, wood, diapers, mixed plastics, with and without PVC, as well as single kinds of plastics like PE, PP, PS, EPS, PA, PET and PVC (Titech, 2011a). However, the separation efficiency is influenced by the existence of black and very dark coloured particles which cannot be separated, as the amount of absorbed NIR light is too high (Hüskens, 2006) and no reflected radiation arrives at the sensor system. Therefore, dependent on the composition of the feed material the recovery only comes up to about 80 to 90%. The attainable product purities reach 90 to 97 wt.-%.

A new field for the application of near infrared sorters is the generation of refuse derived fuel (RDF) which is frequently conducted in plants for the processing of commercial waste or in mechanical-biological waste treatment plants (Nisters, 2006). Generally, sensor based NIR sorters are employed in RDF-processing for the separation of highly calorific constituents like plastics, paper and cardboard, textiles and wood. These materials can, in fact, only be separated very imperfectly with conventional methods like air classification. Moreover, this method allows fading out chlorine and antimony bearing particles (*e.g.* PVC and PET) in the course of the recognition procedure. In this way, it can be prevented that bothersome ingredients are discharged together with the RDF and limit its chlorine content to less than approximately 1 wt.-%.

Figure 5: TITECH NIR separation module in operation.

Fig. **5** shows a NIR separator in operation. It is possible to clearly identify the recognition unit and the IR emitter with halogen lamps, mounted over a belt conveyor covering a width of up to 3.000 mm.

Metal Separation with Inductive Sensors

The rapid increase of prices for scrap metals within the last decade promoted the implementation of sensor based sorting devices equipped with inductive detectors. These devices utilize the great differences of electrical conductivities as separating attributes. For this purpose mostly belt type separators are employed. Fig. **6** illustrates the functioning principle and the construction of sorters equipped with inductive detectors (Titech, 2011b).

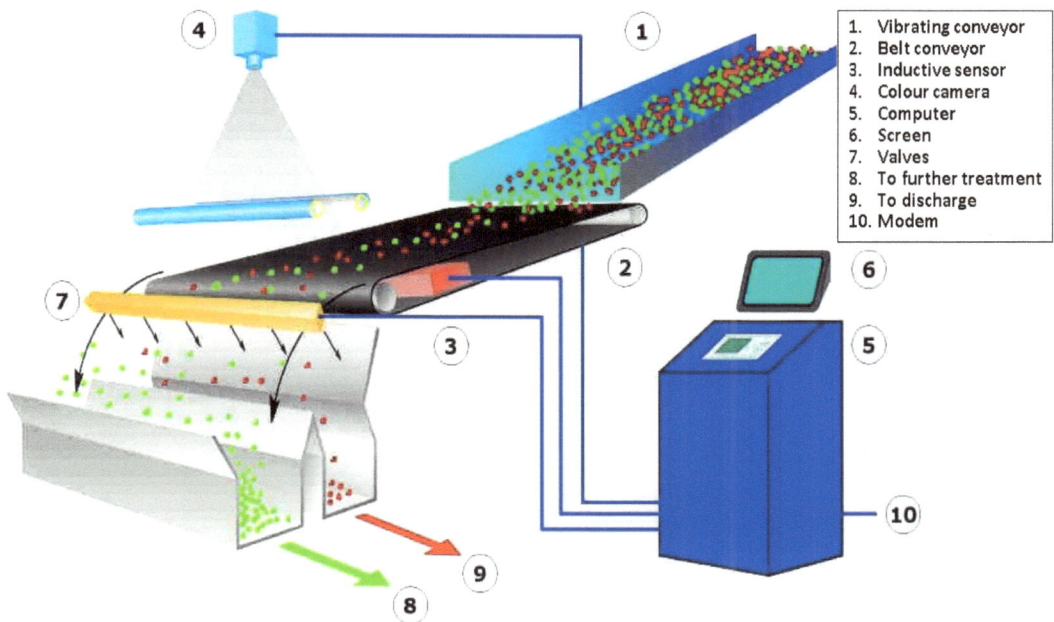

1.	Vibrating conveyor
2.	Belt conveyor
3.	Inductive sensor
4.	Colour camera
5.	Computer
6.	Screen
7.	Valves
8.	To further treatment
9.	To discharge
10.	Modem

Figure 6: Principle of inductive sorter with additional camera sensor ("CombiSense") (Julius and Müller, 2002).

The feed material arrives *via* a vibrating conveyor (1) onto a belt conveyor (2) running with a velocity of 3 m/s. As the strength of a radiated electromagnetic field drops in intensity with the square of the distance, the inductive sensor (3) must be installed as close as possible under the belt. This sensor system is suited for the recognition of all metals as well as for the identification of stainless steel particles.

The additional application of a line scan colour camera (4) has the advantage that position, size and shape of the particles can be identified much better in comparison with a single inductive sensor system (Schunicht, 2007). Finally, the separation is conducted by the activation of valves. Thus, positively recognized particles can be blown out with compressed air blasts from the nozzles (7).

Moreover, this kind of device is capable of separating for instance non-ferrous metal mixtures according to different colours. Thus, it is possible to re-process intermediate goods containing non-ferrous metals to achieve grey (aluminium, zinc, lead), red (copper) and yellow (brass) metal products.

A main field of application for inductive sensor sorters can be found in processing of shredder residues after eddy current separation. This method allows recovering non-ferrous metals and stainless steel which are still contained within these so called tailings. In particular the separation of stainless steel can be greatly improved, as the conventional handpicking method only allows low throughputs and limited separating efficiencies. Experiences during the last decade proved that a multiplicity of metal containing mixtures can be successfully processed with inductive sorting machines.

Fig. 7 shows an example of the assembly of a sensor based metal sorter. The most important technical data are the following:

- Belt conveyor: length 4,000 mm, width 1,200 mm, P = 3 kW;

- Sensors: colour line scan camera with 1 billion pixels, 1 million colours; metal detector with 48 coils;

- Discharge unit: 160 valves, nozzle pattern 8 mm, compressed air supply min. 8 bar and min. 7 m³/min, compressor drive 75 kW;

- Separation of: metals/non metals, stainless steel from non-ferrous metals, colour separation of different mixtures, particle size min. 5 mm;

- Throughput: for metal/non metal mixtures approx. 10 – 20 t/h.

Figure 7: View of TITECH "CombiSense" inductive metal sorter (Titech, internet site (b)).

Density Separation with X-Ray Sensor Devices

Within the last five years automatic sensor sorting devices, operating according to the principle of X-ray transmission measurement, were developed. As it can be seen in Fig. **8**, these devices use an X-ray tube as emitter, which is installed over a fast running belt conveyor. The radiation passes through the components of the feed material as well as through the belt. Two line-scan imaging detectors arranged in the bottom strand of the conveyor are used as sensor system (Harbeck, 2006). These sensors identify the intensity of the X-ray radiation simultaneously with two measuring channels in two different wavelength ranges.

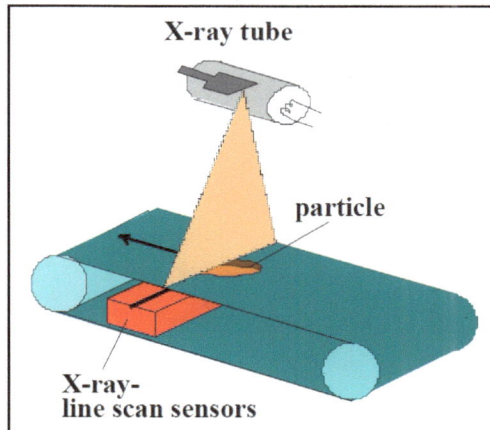

Figure 8: Principle of X-ray -emitter and -sensor arrangement in sorting machines (Harbeck, 2006).

The measures obtained by both channels are evaluated by image processing and suitable algorithms and result in a determination of the particle density. With this method the influence of the particle thickness (2- or 3- dimensional particles) calculation can be eliminated to a great extent.

The utilization of the separating attribute "material density" always allows successful applications if single components of feed mixtures exhibit differences in the specific weight. This kind of processing shows some advantage in comparison with heavy media separation as it is a dry working procedure. Thus, *e.g.* in metal processing, magnesium and aluminium can be separated from heavy metals. Additionally, other typical applications of sensor based separation with X-ray transmission measurement are the selective separation of electronic scrap from light packaging waste as well as the removal of inert materials from construction waste and waste wood.

Fig. **9** gives a view of a separator operating with X-ray transmission detection. With a width of the belt conveyor of 1,200 mm, the throughput of mixed metals, that contains approximately 70 wt.-% of aluminium and 30 wt.-% of heavy metals, comes up to 10 t/h. The separation results proved that an Al-recovery of 93% can be achieved while the aluminium product grade amounted to 97 wt.-% (Harbeck, 2006). The X-ray technology allows in this way to improve the purity of recovered aluminium to a great extent. This is especially due to the great reliability of the X-ray method which guarantees that the separating efficiency is not influenced by any colours or coatings or impurities of the particle surfaces.

Figure 9: View of sensor based sorter with X-ray transmission measurement (Harbeck, 2006).

CONCLUSIONS

In comparison with conventional separation equipment, sensor based sorting opens up completely new applications for dry waste processing. This new field of technology represents a single grain separation which mainly benefited from the progress in the development of new sensors as well as in real-time digital data handling. Thus, the increasing adoption of sensor sorters in waste treatment processes resulted in a considerable innovation permitting to achieve the goal of quality-assured final recycling products.

ACKNOWLEDGEMENT

Declared none.

CONFLICT OF INTEREST

The author(s) confirm that this chapter content has no conflict of interest.

REFERENCES

Dehler, M. (2008). All recyclables from one bin, In: *Proceedings of Conference Sensor Based Sorting*, Department of Processing RWTH Aachen University.

Fears, P. (2008). Maximizing Mechanical Metal Recovery – Eriez ProSort Mechanical Metal sorter, In: *Proceedings of Conference Sensor Based Sorting*, Department of Processing RWTH Aachen University.

Harbeck, H. (2006). Aufbereitung von Aluminiumschrottenmittelssensorgestützter Sortierung, In: *Proceedings of Conference Sensor Based Sorting*, Department of Processing RWTH Aachen University.

Hüskens, J. (2006). SortenreineWertstoffe mit NIR-Nahinfrarot-Technik. *Anwendungen bei der Abfallsortierung, Wasser, Luft, 5,* 58-59.

Julius, J., Müller, J. (2002). Entwicklung und Erprobung eines Sortierverfahrens für die Rückgewinnung der Edelstahlfraktion. *Final report on a development project funded under the reference number 15926 of the German Federal Environmental Foundation.*

Killmann, D., Pretz, T.h. (2006). Status der sensorgestützten Sortierung, In: *Proceedings of Depo-Tech Leoben, Abfall- und Deponietechnik*, Abfallwirtschaft, Altlasten.

Killmann, D., Pretz, T.h. (2007). Perspektiven der sensorgestütztenSortierung, In: *Proceedings of InternationalenTagung MBA*, Hannover.

Mogensen (2004), *Plastic Sorting.* Available from: http://www.mogensen.de/en/kunststoffsortierung.htm [Accessed on 17th February 2012].

Nisters, T.h. (2006). Ersatzbrennstoffherstellungmit NIR- Technologie. *Aufbereitungs-Technik, 47*(12), 28-34.

Pretz, T.h., Julius, J. (2008). Stand der Technik und Entwicklungbei der berührungslosenSortierung von Abfällen. *Österreichische Wasser- und Abfallwirtschaft*, Springer Verlag, Wien/New York, 07-08, 105-112.

Schunicht, J. (2007). Rückgewinnung von Metallenaus E-Schrott – ModerneSortierverfahren. *Praxis, Aufbereitungs-Technik, 48*(9), 4-10.

Titech. (2011a). *Brochure of Titech Autosort.* Available from: http://www.titech.com/sorting-equipment/titech-autosort-10715[Accessed on 17th February 2012].

Titech. (2011b). *Brochure of Titech Combisense.* Available from: http://www.titech.com/sorting-equipment/titech-combisense-10717 [Accessed on 17th February 2012].

Zeiger, E. (2004). Sortierung von PET-Flakes. *Sekundär-Rohstoffe*, 3, 80-82.

Zeiger, E. (2005). Glasrecyclingmit Mogensen Sortier- und Siebtechnik. *Aufbereitungs-Technik, 46*(6), 6-13.

Send Orders of Reprints at reprints@benthamscience.org
Separating Pro-Environment, 2012, 77-89

CHAPTER 5

Upgrading of Post-Consumer Steel Scrap

F. Quarta[1], A. Bonoli[1] and P.C. Rem[2,*]

[1]*Department of Civil Environment Materials Engineering, University of Bologna, Italy and* [2]*Department of Resources & Recycling, Delft University of Technology, The Netherlands*

Abstract: Post-consumer steel scrap resulting from End-of-Life Vehicles (ELV), Waste from Electric and Electronic Equipment (WEEE) and Incineration Bottom Ash (IBA) is often hand picked for copper, stone, cloth and other contaminants, in order to meet the specifications of the steelmakers. At capacities of 20 t/h of scrap or more, the efficiency of hand sorting generally becomes problematic, hence, leaving half of the copper in the steel product. New technologies are presently being proposed to facilitate or even eliminate hand sorting of these types of scrap, allowing operators to increase revenues from copper, expand plant capacity, realize a higher and more consistent steel product quality or avoid legal constraints associated with hand pickers. A shape-sensitive magnetic separator called "Clean Scrap Machine" (CSM) pre-sorts the scrap into a bulky thin-walled steel fraction of consistently high purity and a volumetrically much smaller flow of relatively heavy parts in which the contaminants are concentrated. The latter flow can either be sorted by a much smaller number of hand pickers, by sensor sorting, or it can be sold directly to specialized sorters to extract the copper. Detailed results are reported for mid-sized IBA scrap.

Keywords: Steel scrap, Recycling, Hand picking.

INTRODUCTION

In 2007, 1.3 billion metric tons of steel produced worldwide consumed about half a billion tons of steel scrap (IISI – International Iron and Steel Institute - data, 2008). Despite this impressive volume of recycled scrap in absolute numbers, the relative contribution of scrap to the production of new steel (about 37%) was actually historically low in 2007, as a result of the steep increase of new steel production in the fast-growing Far East economies. Figures for Germany (Hüther, 2010) show a 45% contribution of scrap to new steel, suggesting that a higher

*Address correspondence to P.C. Rem:** Department of Resources & Recycling, Delft University of Technology, Postbus 5048, 2600 GA, Delft, The Netherlands; Tel: +31(0)1527 83617; Fax: +31(0)1527 88162; E-mail: P.C.Rem@tudelft.nl

input of scrap is possible in stabilized economies. From the perspective of the environment, using more scrap is a positive development. The Fraunhofer Institute for Environmental, Safety and Energy Technology (2008) reports 0.68 t of CO_2 emissions for a ton of recycled steel *versus* 1.54 t of CO_2 for a ton of primary produced steel. Yet, there are also problems with maximizing the recycling of steel. About 10%, or 50 million tons, of the steel scrap that becomes available worldwide annually is post-consumer scrap that is contaminated with elements such as copper, tin, zinc, chromium, nickel, molybdenum, phosphorus and sulfur. This scrap results from End-of-Life Vehicles (ELV), Waste from Electric and Electronic Equipment (WEEE) and Municipal Solid Waste (MSW) Incineration Bottom Ash (IBA) (see Table **1**).

Table 1: EU Statistics of post-consumer scrap

Type of steel scrap	EU production (Mt/y)	Typical capacities[*] (t/h)	Typical contaminants/levels
WEEE	2	1-5	2.3% Cu
IBA	1	2-40	0.7% Cu, Phosphor, 0.1% S, sand, 0.2% coarse stone, 0.1% cloth/plastic
ELV	8-11	30-200	0,7% Cu, rubber, stainless, cast Al
Total	11-14		

[*]Input steel scrap for hand sorting operations.

Contaminating elements like copper and tin affect the mechanical strength and resistance to corrosion of the steel product (Savov *et al.*, 2003), whereas phosphor and sulfur pose problems during smelting. Depending on the type of smelter (electric arc furnace or blast furnace) and the required steel quality (from rebar to cold rolled steel), maximum contaminant concentrations in the scrap vary (*e.g.* for Cu: between 0.04% and 0.4%). Unfortunately, some elements also build up in the steel matrix, increasing the need for scrap purification with each life cycle in countries with high rate of scrap recycling.

Widely implemented solutions for reducing the contamination levels of post-consumer scrap are mechanical processing and hand picking. Combinations of screening, shredding and magnetic separation are able to concentrate copper, stainless steel, cast aluminum, stone, dirt and cloth into fine, non-magnetic or weakly magnetic fractions. Hand-sorting primarily serves to reduce the content of relatively large pieces of stone and cloth, and recover the valuable fraction of copper-containing parts from the scrap, consisting mainly of electrical motors, transformers and electric wires. Analysis of the existing purification technologies shows that there is still a lot to gain, both in terms of process costs and in terms of value recovery. Shredding the steel scrap to liberate the contaminants so that they can be separated from the scrap by magnets is relatively expensive (about 25 €/t of scrap) and not fully effective. The cost of hand-sorting is lower, typically 5 €/t of scrap, but the efficiency of contaminant removal depends strongly on the capacity. For flows of WEEE scrap of 5 t/h, a well-managed hand-sorting team can recover 90% of the copper content, but at 20 t/h or more the efficiency drops to 50% or less. This means that, at typical copper contents in IBA and automotive scrap of between 0.4% and 0.7%, from 12 € to 17 € of copper scrap value is left in each ton of scrap, even apart from the problematic quality of the steel scrap itself. Another problem of hand-sorting is that the efficiency of contaminant removal is fluctuating, difficult to control, and that some contaminants, such as stainless steel or cast aluminum are not recognized by hand-pickers. A final issue is that the legislation of some countries explicitly discourages the use of hand-pickers.

Figure 1: Side view (left) and front view (right) of the shape-sensitive magnetic separator of Resteel.

AUTOMATIC UPGRADING OF STEEL SCRAP

From a technical point of view, an obvious alternative to hand sorting is sensor sorting. In principle, sensors, or combinations of sensors, are able to identify contaminant particles in the scrap with a similar precision as human sorters, and the use of sensors circumvents the problems of quality control, inconsistent product purity, contaminants that are difficult to distinguish from steel by the naked eye, and legislation. However the problem of process cost and limited capacity remain. Sensor sorting is relatively expensive and it is extremely difficult to remove contaminant particles from steel scrap at high throughputs.

To solve the problems of cost and capacity, a technology called CSM (Fig. **1**) was developed by Delft University of Technology (Rem *et al.*, 2010) to first eliminate steel scrap particles with a flat or longitudinal shape from the scrap by a shape-sensitive magnetic field, before sensor sorting or hand picking. To this end, a magnet is introduced into the pulley of a conveyor belt with field lines that are almost parallel to the curved part of the belt surface where the scrap particles follow different trajectories according to the difference between magnetic and centrifugal forces (Fig. **2**). Flat and longitudinal steel particles (Fig. **3**) are

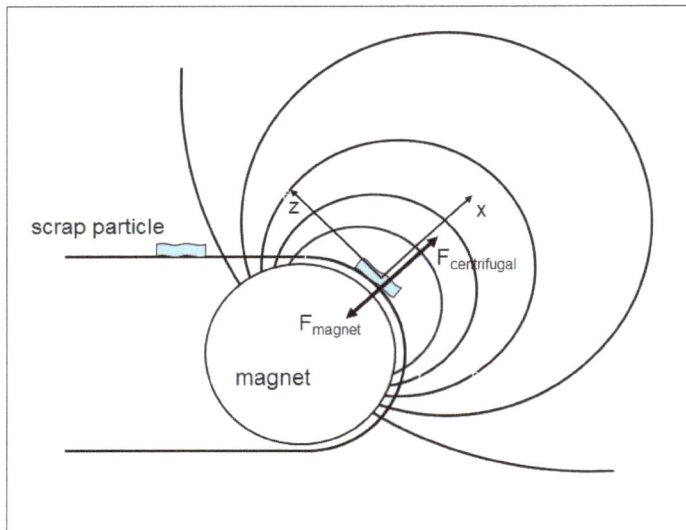

Figure 2: Side view of the pulley, magnet and magnetic field of the CSM technology.

strongly attracted by such a magnetic field, even when the field intensity is relatively weak, whereas more compact steel particles and non-magnetic materials, such as stones and cloth (Fig. **4**) are more weakly attracted, or not at all. Electrical motors, transformers, copper wires, stones, cloth, stainless, cast aluminum and rubber parts with steel inserts all belong to the latter category, and are therefore released from the belt at an early point. The flat and elongated steel pieces move with the belt and are separated from the rest of the scrap. The shape-sensitivity of the CSM magnetic field may be understood qualitatively from the force balance for particles marginally following the belt surface at the point of separation (Fig. **2**),

$$F_{centrifugal} = F_{magnet} \tag{1}$$

Figure 3: Flat and longitudinal steel particles reporting to the clean scrap product.

Figure 4: Compact steel parts and contaminants reporting to the contaminant concentrate.

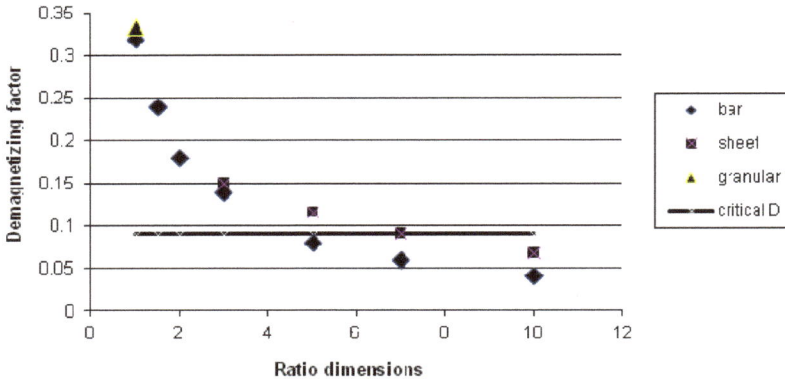

Figure 5: Demagnetizing factor for ferromagnetic parts of simple shapes, as a function of the ratio of the largest dimension to the smallest dimension.

Evaluating both sides of the equation for a steel particle of mass m and volume V, moving with the velocity v of the conveyor at radius R around the magnet, we get:

$$\frac{V}{D_z} H_z \left| \frac{\partial B_z}{\partial x} \right| = \frac{mv^2}{R} \tag{2}$$

In this equation, H_z and $B_z = \mu_0 H_z$ are the magnetic field and magnetic induction, respectively, and D_z is the demagnetizing (shape) factor of the steel part (Hopstock, 1985). The magnetic induction of the CSM magnet at the particle position on the belt surface is $B_0 \approx 0.08$ Tesla, and its gradient is given by:

$$\left| \frac{\partial B_z}{\partial x} \right| = \frac{2B_0}{R} \tag{3}$$

This means that the critical demagnetizing factor of the steel particle for which the centrifugal force just matches the magnet force is (ρ is the density of steel):

$$D_z = \frac{2B_0^2}{\mu_0 \rho v^2}; \quad \rho = m/V \tag{4}$$

Inserting typical values, $\rho = 8000$ kg/m^3, $\mu_0 = 4\pi\ 10^{-7}$ Tesla m/A, and $v = 4$m/s, the critical value for the demagnetizing factor for a steel particle to remain attracted to the drum is about 0.09. Fig. **5** shows the demagnetizing factors for simple steel bars and flat pieces of varying dimension ratios. It is clear that roughly spherical particle shapes have a demagnetizing factor higher than 0.09, while bar-shaped and flat pieces have lower demagnetizing factors. Actual shapes of scrap particles are often more complex than simple bars and sheets, and their motion at the point of separation depends on more factors than the magnetic-centrifugal force balance alone. Yet, low demagnetizing factors are normal for clean steel pieces in post-consumer scrap, while high demagnetizing factors are typical for magnetic contaminant parts. Fig. **6** shows the statistics of the demagnetizing factor for particles from IBA scrap, indicating that 85 mass% of the clean steel can be recovered in the clean scrap product while 90% of the copper-containing parts report to the contaminant concentrate.

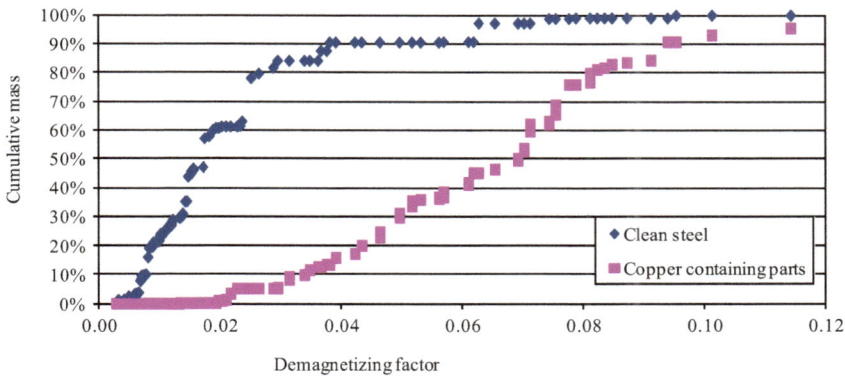

Figure 6: Cumulative mass distribution of the measured demagnetizing factor of over a hundred clean steel particles (diamonds) and contaminated ferromagnetic scrap particles (squares) randomly selected from IBA scrap.

The amount of steel scrap that can be separated by CSM as a clean product (assuming a 0.1% Cu metal content for the clean product) depends on the type of scrap, and varies between about 60 mass% of the input for some automotive scrap and 85 mass% for mid-size IBA scrap. The amount to be sorted by hand or sensors therefore reduces to 15% - 40% of the input. However, since the clean scrap product contains the light and bulky flat and elongated pieces, the volume (Table **2**), the belt coverage (Fig. **7**) and number of particles of the contaminant concentrate are reduced more strongly than suggested by the mass split, and it is volume, belt coverage and particle number which are critical for sensor sorting or hand sorting capacity. In 2009, a Dutch upgrading plant for IBA scrap replaced its hand-sorting operation of 20 t/h using a team of eight hand-pickers by a CSM running at 40 t/h and a team of four hand-pickers to clean the contaminant concentrate of the CSM. In this way, the plant was de-bottlenecked and the cost of hand sorting a ton of scrap was reduced by a factor of four, despite the fact that the number of contaminant parts to be hand-picked remained the same.

The wall thickness of steel scrap is an important parameter in feeding the scrap to the smelt, both for electric arc furnaces and blast furnaces. Thick-walled, high bulk density scrap behaves differently in the smelt than thin-walled scrap, and smelters, therefore, schedule the feeding according to wall thickness. Since the clean scrap product of the CSM consists largely of thin sheet and has a lower bulk density than the steel scrap produced from the contaminant concentrate by hand-sorting, an interesting option is to market the two scrap products separately.

Table 2: Statistics of example cases of CSM separations on post-consumer scrap.

Type of Steel Scrap	Clean Scrap Product			Contaminant Concentrate		
	Mass [%]	Bulk density [t/m^3]	Volume [%]	Mass [%]	Bulk density [t/m^3]	Volume [%]
WEEE	73*	0.7	78	27	0.9	22
IBA	72*	0.9	85	28	2	15
ELV	75*	0.7	85	25	1.3	15

* Primary variable dependent on machine settings: range is 60% - 85%.

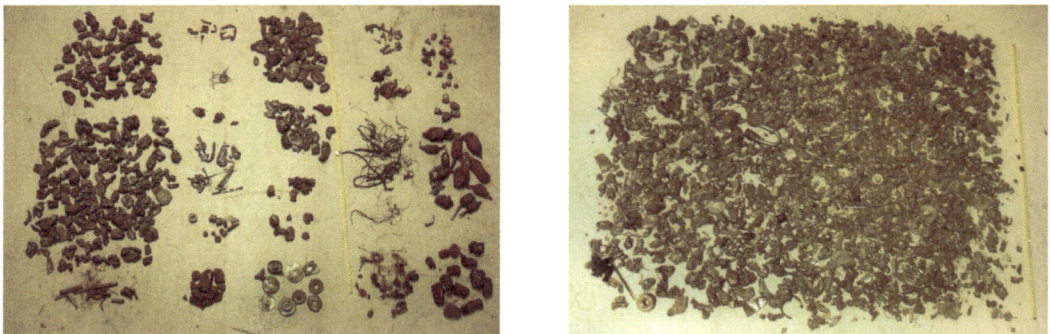

Figure 7: Clean scrap product (left) and contaminant concentrate (right) for a batch of automotive scrap, spread out on a flat surface to show the improved presentation of the copper contaminant particles (right from the ruler at top) for sensor or hand sorting.

Table 3: Parameters used in waste fraction definition of IBA scrap

Parameter	Range
Material (combination)	batteries, plastics, mineral, stainless, cast iron, wire steel, other steel
Thickness	<1 mm, 1–2 mm, >2 mm
Shape	sheet, folded sheet, rod, folded wire, granular, tube
Screen size	5-15 mm, 15-30 mm, 30-45 mm, 45-70 mm, >70 mm

WASTE FRACTION ANALYSIS FOR IBA SCRAP

Models of shape-sensitive magnetic separation that are based on first principles, such as the simple one presented in the previous section, are not very useful for process engineering. Such models cannot deal with the complex shape of actual scrap particles, and even if they could, it would be very expensive to feed detailed particle characteristics to the model. In order to model separators like the CSM for the purpose of engineering, an alternative approach based on waste fraction analysis was tried. The concept of this approach is to first divide the scrap into a large number of fractions, each fraction containing particles of the same screen size classes, material (or combination of materials) and shape/type of EOL product. The recovery of each fraction into the clean steel product is then measured for a series of alternative cut-points of the separator. From these results, the separation results for a new batch of scrap can be predicted by analyzing this scrap in terms of the same waste fraction definitions and assuming that the results for each waste fraction is the same as for the original batch.

For the present study, tests have been carried out on seven materials or combinations of materials that are typically present in IBA scrap. A batch of 500 kg of mid-size IBA scrap was screened at 5 mm, 15 mm, 30 mm, 45 mm and 70 mm to produce six size classes. Since the <5 mm class consisted mainly of sand, only the remaining five size classes were hand sorted for the seven materials. Finally, each resulting fraction was cut according to six particle shapes and three material thickness ranges (see Table **3**). By means of this characterization, a large number (110) of waste fractions were obtained, each waste fraction having scrap particles of a specific material m, thickness g, shape g' and size class g''. Tests with the CSM were set up with 11 splitters at increasing distances from the separation point so as to create 12 collection bins, $x = 1,.,12$. Each of the waste fractions was passed over the machine separate from the others, and the mass distribution over the collection bins was documented for each fraction. It was found that the distributions of particles over the product bins were quite narrow, as it was expected because of the strong similarity of the particles within a fraction. So the result of the separation was described by the mean collection position (average bin sequence number) $x_{m,g,g',g''}$ of each fraction.

Since it is too much time consuming to characterize a batch of scrap according to four different particle parameters, it was investigated to which extent the four parameters played a significant role in the prediction of the separation result. Five models were compared (see Table **4**):

- **Model I:** The predicted position is the same for all particles, irrespective of material, particle dimensions or particle shape; it is the mass-averaged position of the batch;

- **Model II:** The predicted position is the same for all particles of the same material m; it is the mass-averaged position of all particles of that material;

- **Models III, IV and V:** The predicted position is the mass-averaged position of all particles of material m having property g, g' or g''.

The basis for comparison of the models was chosen as:

$$\text{TV} = \sum_m \sum_g \sum_{g'} \sum_{g''} \frac{M_{m,g,g',g''}}{M_m} \left(x_{m,g,g',g''} - X_{m,g,g',g''} \right)^2 \tag{5}$$

where $X_{m,g,g',g''}$ is the predicted position of the particle of fraction m,g,g',g'' according to the model and $M_{m,g,g',g''}$ and M_m are the mass of the fraction and the mass of material m in the batch, respectively. This formula for the Total Variance (TV) was selected because it assigns the same importance to each material and weighs sub-fractions of a given material according to their mass contribution to the batch.

If the predicted position is chosen to be the same for all waste fractions (Model I), the resulting TV is found to be 32, meaning that, on average for the seven materials, the prediction of particle positions by the model is out by 2.2 product bins. Model II, predicting separate positions for each of the different materials (see Table **4**), reduces the TV to 13. From the Analysis of Variance (ANOVA) point of view, this model adds six estimated parameters (seven predicted positions instead of one), so that the reduction of the total variance per added parameter is about 3. Since this is much higher than the original total variance per waste fraction (degree of freedom), which is about 0.3, the reduction is statistically significant. On average, the error in predicting the position of a scrap particle is about 1.4 product bin.

Table 4: Predicted scrap particle positions (averaged sequence number of collection bin) according to alternative models

Model		Batteries	Plastics	Minerals	Stainless	Cast iron	Wire Steel	Other Steel
II		5.3	6.2	7.3	3.0	4.7	1.8	2.7
III	<1 mm	5.3	1.9				1.4	2.6
	1-2 mm		6.1		3.2		1.8	2.3
	>2 mm		7.2	7.3	2.6	4.7	1.8	4.1
IV	Sheet		6.9		2.9	3.7		2.8
	Folded sheet		5.5			3.0		2.0
	Rod					5.9	1.8	
	Folded wire						1.8	
	Granular			7.3				
	Tube	5.3			3.9			4.1
V	5-15 mm							1.8
	15-30 mm	5.8		6.9	3.7		2	1.8
	30-45 mm	4	6.5	7.4	3.0	5.4	1.7	2.0
	45-70 mm		5.4		1.5	4.5	1.6	2.1
	>70 mm		7	7.2		4.6	1.8	3.4

Model III (prediction on the basis of material and particle thickness, 14 parameters) adds another seven parameters but reduces the total variance to 4.6, a reduction of 1.2 for each added parameter, and it reduces the prediction error to an average of 0.8 product bin. Again, this reduction is about ten times what would be expected for a non-physical model, and therefore a statistically significant result. Models IV and V, based on classifications of the waste according to material and particle shape or size, result in a total variance of 5.6 and 6.5, respectively. Since the classification on the basis of size adds 16 extra model parameters, particle size is a less important parameter than particle thickness. Particle shape, adding seven extra parameters, is just secondary to particle thickness.

The results of waste fraction analysis suggest that models for shape-sensitive magnetic separation that correlate particle trajectories with particle material and thickness are reasonably good predictors of separation performance. This is in fact in agreement with the large observed differences in the bulk densities of the clean steel product and the contaminant concentrate.

CONCLUSIONS

There exists a strong environmental drive to maximize the input of steel scrap in the production of new steel. Yet, in order to produce high quality steel, the contaminant levels in post-consumer scrap must be reduced, by hand picking or by sensor sorting. Both of these options currently have a problem with cost and high capacities of 20 t/h or more. An interesting possibility is to use shape-sensitive magnetic sorting to remove the bulky but clean flat and rod-shaped parts from the scrap, prior to sorting out the contaminants. Results from a CSM-retrofitted Dutch processing site for IBA scrap show an increase in capacity of the scrap treatment line from 20 t/h to 40 t/h and a reduction of hand sorting costs per ton by a factor of 4. A positive side effect of using shape-sensitive magnetic sorting is that two steel scrap products are produced of strongly different bulk densities, offering the possibility to optimize feeding of the clean scrap to the smelt. A reasonably predictive model for the separation performance of shape-sensitive magnetic sorting on scrap streams can be based on correlations with particle material and thickness.

ACKNOWLEDGEMENT

Declared none.

CONFLICT OF INTEREST

The author(s) confirm that this chapter content has no conflict of interest.

REFERENCES

Fraunhofer Institute for Environmental, Safety and Energy Technology (2008). Recycling delivers 'stunning' emission savings. *Recycling International*, *8*, 44-49.

Hopstock, D.M. (1985). Magnetic Forces. *SME Mineral processing handbook*, N.L. Weiss editor, Society of Mining Engineers, New York.

Hüther, M. (2010). *Volkswirtschaftliche Bedeutung der Entsorgungs- und Rohstoffwirtschaft*, Institut der deutschen Wirtschaft Köln.

International Iron and Steel Institute (2008). *Fact sheet raw materials*. Available at: www.worldsteel.org and http://elibrary.steel.org.au [Accessed 17th February 2012].

Rem, P.C., Berkhout, S.P.M., Fraunholcz, O.N. (2010). *Process and device for the separation of fragments of liberated ferrous scrap from not liberated ferrous scrap fragments by means of a static magnet*, European Patent 2206558.

Savov, L., Volkova, E., Janke, D. (2003). Copper and Tin in Steel Scrap Recycling. *RMZ - Materials and Geoenvironment*, *50*(3), 627-640.

CHAPTER 6

State of the Art of Recycling of Photovoltaic Panels Using Separation Technology

A. Bonoli* and A. Pompei

Department of Civil Environment Materials Engineering, University of Bologna, Italy

Abstract: The exponential growth of photovoltaic (PV) installations highlights the necessity to cope with the environmental impacts which could raise from wrong practices for disposal of end-of-life PV modules. In fact, their possible disposal in landfills would represent a loss of materials and energy, as PV modules are goods that can become very useful even at the end of their life.

In order to improve the valorisation of waste coming from end-of-life PV modules and of materials they contain, high value recycling processes are needed. These solutions differ from other end-of-life management alternatives, that are generally more simple and economical, like for example the treatment of the PV modules in a recycling plant for laminated glass or their disposal at a landfill after the recovery of the aluminium frame and a pre-treatment in a municipal incineration plant. All these low value approaches have in common the loss of valuable resources, represented by high value materials and components contained in PV modules (wafer/silicon, indium, tellurium and so on), which inevitably would end their life cycle in a landfill, without being adequately recovered.

Differently, some separation and treatment processes have been recently developed, adopting machines and also plants deriving from traditional mineral processing, which allow recovering and recycling of precious metals and materials.

Keywords: Silica, Metals recovery, Recycling, Separation, PV modules.

INTRODUCTION

As it is well known, solar energy has electromagnetic origin: in fact radiant energy, coming from solar radiations, can be directly converted into useful energy (thermal or electrical). In the case of photovoltaic installations, solar radiation is

*Address correspondence to A. Bonoli: Department of Civil Environment Materials Engineering, University of Bologna, Via Terracini, 28, Bologna, Italy; Tel.: +390512090234; Fax: +390512090308; Email: alessandra.bonoli@unibo.it

converted into electrical energy, as hitting the semiconductor contained in the PV modules it triggers a flow of electrons and, consequently, a flow of electricity through it (Bianchi *et al.*, 2004). Solar energy is an alternative and inexhaustible source of energy and, at the same time, it is helps protecting the environment as, during the operating life of a photovoltaic plant, there are no atmospheric emissions of pollutants (like for example NO_X and SO_X) and of the so called "greenhouse gases" (CO_2), that contribute to climate change and global warming. Therefore, photovoltaic energy production shows unquestionable environmental benefits compared with typical generation systems based on the use of fossil fuels. However, the exponential growth of photovoltaic installations (Gabrielli, 2010) is forcing PV module producers to cope with the possible environmental impacts coming from wrong practices for disposal of end-of-life PV modules.

Until a few years ago, the issue of the disposal of PV panels was of little importance because photovoltaic technology was quite new and because of the long life of PV modules, which is estimated to be at least 25-30 years. In recent years, the significant expansion of photovoltaic installations in Germany, USA and, to a lesser extent, in Spain and other European countries, makes it clear the urgent need to find proper solutions to ensure a sustainable disposal of end-of-life PV modules.

The disposal of PV modules in landfills would represent a loss of materials and energy, as PV modules are goods that can be very useful also at the end of their life cycle. A typical crystalline PV module, for example, is composed of:

- Glass, which represents about 70% of its weight, used for the protective and exposed surface of the module;

- Metals, *e.g.* aluminium, used for the frames;

- Silicon, the semiconductor material; and

- Other materials, like for example silver and copper used for the electrical contacts of solar cells.

These materials are valuable and their life cycles can be different from those of photovoltaic devices containing them. Such materials could be therefore recovered or recycled in view of a possible further use in the production of new PV modules or of other products. Moreover, it is of fundamental importance to characterize PV modules in order to classify them as hazardous wastes or not on the basis of their content of toxic metals (Cd, Pb, *etc.*). For hazardous wastes, in fact, it is necessary to respect more strictly handling, disposal, record keeping and reporting requirements, in order to ensure a high level of protection of the environment and human health.

In this regard, the U.S. Environmental Protection Agency (EPA) has defined some specific procedures and laboratory tests to classify waste, therefore also those waste coming from dismissed PV modules (Fthenakis e Eberspacher, 1997). Should PV modules fall under the category of "hazardous wastes", their recycling would represent an advantageous alternative, not only from an ethical and ecological point of view, but also economic.

TYPES OF SOLAR CELLS AND OF SEPARABLE MATERIALS

The materials mainly used in the production of PV modules are: glass, aluminium, EVA (Ethylene Vinyl Acetate), silicon, copper, silver, tin and lead.

Tables **1** and **2** show the average composition of the main typologies of PV modules (figures 2007).

Table 1: Average composition of a current c-Si standard module – Source: Sander *et al.*, 2007

Component	Quantity (2003)	Quantities (2007)	
	Weight (%)	Weight (%)	(kg/kW$_p$)
Glass	62.7	74.16	77.3
Frame Aluminium	22	10.3	10.7
EVA	7.5	6.55	6.8
Solar Cells	4	3.48	3.6
Backing Film (Tedlar)	2.5	3.6	3.8
Junction Box	1.2	-	-
Adhesive potting compound	no data	1.16	1.2
Weight/kW$_p$	103.6 kg/kW$_p$	-	103.4
Cu	0.37	0.57	-
Ag	0.14	0.004-0.006	-
Sn	0.12	0.12	-
Pb	0.12	0.07	-
Si	no data	3	-

Silicon Solar Cells, Crystalline (Monocrystalline or Polycrystalline) and Amorphous

In PV modules that make use of slices of thin wafers of crystalline silicon (c-Si) (Table **1**) the materials mainly used are glass, aluminium (for the frames) and EVA (Ethylene Vinyl Acetate). As regards the PV modules in which amorphous silicon is used (Table **2**), crystalline silicon wafers are replaced by a thinnest layer (some microns) of semiconductor material (amorphous silicon), which is directly deposited on a glass substrate.

Table 2: Average composition of an a-Si module with frame in polyurethane – Source: Sander *et al.*, 2007

Material	Thickness	Weight per Module	Weight per Surface	Weight per Output
		(g)	(g/m^2)	(g/W$_p$)
Glass	2.2-3 mm	3,483	12,480	249.6
SnO$_2$		0.96	3.45	0.069
Tin (as oxide)	about 500 nm	0.76	2.72	0.0544857
Boron		1.18E-05	4.23E-05	8.46E-07
Silicon	about 400 nm	0.26	0.92	0.0184
Phosphorus		1.21E-07	4.33E-07	8.66E-09
Aluminium	< 600 nm	0.452	1.62	0.032
Aluminium Strips	0.05 mm	0.988	3.54	0.07
Acryl Resin	0.15 mm	19	68	1.63
Hot melt glue		0.8	2.87	0.057
Cable		40	143	2.86
Polyol		285	1,021	20.3
MDI		215	770	15.4
Total		4,046	14,497	290

CIS (Copper Indium Diselenide)/CIGS (Copper Indium Gallium Diselenide) Solar Cells

In these types of PV modules, the semiconductor material is a polycrystalline compound, sometimes enriched with gallium in order to increase performances, usually deposited on a glass substrate. Typically for this kind of solar cells, the adopted compound is copper indium gallium diselenide (Cu(In,Ga)Se$_2$, CIGS) or copper indium diselenide (CuInSe$_2$, CIS), if gallium is not present.

The typical efficiency of these modules is included in a range which falls between 8 and 10%.

CdTe Solar Cells

The semiconductor material is a polycrystalline compound called cadmium telluride (Table **3**), which is deposited on a glass substrate adopting about the same technologies used for CIS modules.

Table 3: Average composition of a typical CdTe module (120 cm x 60 cm)

Material	Thickness	Weight per Module	Weight per Surface	Weight per Output
		(g)	(g/m²)	(g/W$_p$)
Substrate Glass	3 mm	5,400	7,500	71
TCO-SnO$_2$	<1 μm	4.968	6.9	0.07
CdS	<0.1 μm	0.34776	0.483	0.005
CdTe	7 μm	12.96	18	0.17
Back Contact Metal	1 μm	1.944	2.7	0.03
EVA	0.5 μm	360	500	4.8
Front Glass	3 mm	5400	7500	71
CuSn band		4.9968	6.94	0.07
Junction Box		15.624	21.7	0.2
Cable		41.616	57.8	0.6
Frame		2,160	3,000	28.8
Total		13,402.8	18,615	178.704

Only polymeric materials, like EVA and tedlar, used for the backing film, represent the non-recoverable fractions and can be disposed of in landfill sites or for energy recovery. The glass is partially reusable (approximately 90%) because of possible damages and breakages, and partially recyclable, as the broken parts could be treated in smelters which make use of silica for their fluxing operations. The aluminium frames can be completely recovered and reused, or recycled in the aluminium industry, as it already happens for this material. Other metals, like silver and aluminium, which are used for electrical contacts of solar cells, and copper, used for cables, can be recovered as secondary raw materials. Furthermore when silicon solar cells are undamaged recovered, they can be reused in order to manufacture new recycled PV modules, and consequently saving energy and raw material (silicon). In the case of CdTe modules, the recovery and recycling of cadmium, highly toxic material, reusable in the production of new CdTe modules, would reduce the environmental concerns linked to the presence of this material in PV modules.

HIGH VALUE RECYCLING SOLUTIONS

In order to improve the valorisation of waste coming from end-of-life PV modules and of the resources they contain, high value recycling processes are needed.

These solutions differ from other end-of-life management alternatives, that are generally more simple and economic, like for example the treatment of the PV modules in a recycling plant for laminated glass or their disposal at a landfill after the recovery of the aluminium frame and a pre-treatment in a municipal incineration plant. All these low value approaches have in common the loss of valuable resources, represented by the high value materials and components contained in the PV modules (wafer/silicon, indium, tellurium and so on), which inevitably would end their life cycle in a landfill, without being adequately recovered.

Differently, some PV module producers have recently developed separation and treatment processes, that can be considered as high value recycling processes, notably, the treatment process of First Solar, used for CdTe modules, and the treatment process of Deutsche Solar (subsidiary of SolarWorld), mainly used for crystalline silicon modules. Further, for the treatment of other typologies of thin film technologies the following recycling processes that are still in phase of development, and that give a very important role to physical-mechanical separation, can be considered as high value ones: the RESOLVED process (Recovery of Solar Valuable Materials, Enrichment and Decontamination) and the SENSE process (Sustainability Evaluation of Solar Energy System).

First Solar's Recycling Process

First Solar, leader in the production of CdTe modules, has developed a high value recycling solution for the end-of-life management of its PV modules, mainly based on the use of separation technologies (First Solar, 2010; Sander *et al.*, 2007).

Fig. **1** shows the layout of the recycling plant. Each step of the process is described more in detail in the following.

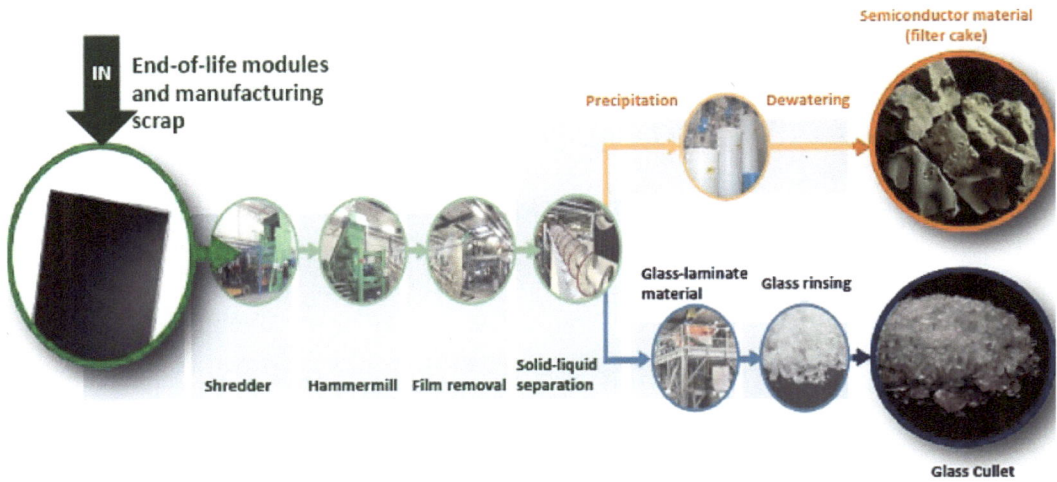

Figure 1: Recycling process of First Solar (CdTe modules).

Size Reduction

The modules are reduced in size in a two step process. The first step utilizes a shredder to break the modules in large pieces, about 10-20 cm, which are then transported by a closed conveyor to the secondary comminution step. In this second step a hammer mill has been adopted in order to further reduce in size the glass, which represents the main fraction of the comminuted PV modules.

In fact, the glass must be crushed into approximately 4-5 mm pieces, in order to ensure the breakage of the lamination bond created by the lamination process. Then these pieces are transported by another closed conveyor towards the next step. Thanks to size reduction, it is possible to recycle, with this method, both broken and undamaged modules.

Semiconductor Film Removal

The comminuted modules are put in a slow rating leach drum made of stainless steel, over a period up to 4-6 hours. Weak sulphuric acid and hydrogen peroxide solutions are added in the drum until an optimal solid-liquid ratio is achieved. In this way, it is put in place a leaching process which allows removing by etching semiconductor films. At the end of the leach cycle, the rotation of the drum is reversed to empty out the contents.

Solid-Liquid Separation

The contents of the leach drum are slowly poured into a classifier including a screw conveyor in order to separate the glass from the liquids. In fact, the screw uplifts the solid fraction, mainly composed by glass and EVA, while liquid is left on the bottom of the classifier and then drained towards the precipitation unit. From this point on, the process splits into two separate paths: the first one for the liquid fraction with high metal content and the second one for the solid fraction.

Glass-EVA Separation

The glass is separated from the encapsulant (EVA), used to join the two glass plates forming the PV module, by a vibrating screen. The pieces of EVA, being of larger size, are transported on the top of the screen by the vibrating action, and then collected on a small conveyor and, finally, properly discharged.

The glass fraction, being of smaller size, falls through the screen. In this way, it is possible to collect the glass at the bottom of the screen and it sent to the successive rinsing step.

Glass Rinsing

The glass coming from the screen separation step is deposited on a tightly woven belt. While being transported on the belt, the glass is rinsed. The water, used for cleaning, removes possible semiconductor residues from the glass and then comes out from the underside of the belt. The clean glass is packed and sent to recycling, while the rinse water is pumped to the precipitation system for metals recovery.

Precipitation

The liquid fractions with a high metal content, coming from the solid-liquid separation and glass rinsing steps, are pumped to the precipitation unit. So the metal compounds are precipitated in a three-step process with increasing values of pH.

Dewatering

Subsequently, the precipitated materials (a sort of sludge) are concentrated in a thickening tank where the solids are allowed to settle at the bottom, while the clean water is pumped off. The thickened material is pumped to a plate and frame filter press for dewatering. Finally, the resulting filter cake, with a high content of metals (cadmium and tellurium), is packed and sent to an external supplier, in order to be refined and reused in the production of the semiconductor material for new PV modules.

In order to control the production and the emissions of dust in all the dry steps of the recycling process, like for example the size reduction one, it is adopted an aspiration system, equipped with a high efficiency particulate air (HEPA) filter, which allows to capture airborne solid particles.

The separation efficiencies of the above described process are very high: it is possible to recover 90% in weight of glass to be used in the production of new glass products and 95% of the semiconductor material to be used in the production of new PV modules. This is particularly interesting because tellurium, used in the photoactive layer, is a relatively rare metal, whose price has been considerably increasing in the last years. As it is expected that thin films market share will increase quickly, tellurium demand will increase as consequence.

The process described above allows recovering semiconductor materials, and therefore it contributes making the thin film technology, based on cadmium telluride, really sustainable and reducing the environmental concerns linked to the use of cadmium in PV modules.

Moreover, without recycling, the development and the diffusion of this technology could be limited in future by a possible depletion of tellurium reserves.

Deutsche Solar's Recycling Process

Deutsche Solar has developed a chemical and thermal recycling process (Fig. 2) for the treatment of the most common crystalline silicon PV modules. A pilot plant to carry out this recycling process has been recently completed with an automation system for the separation step, replacing the manual separation.

As shown in Fig. 2, Deutsche Solar's recycling solution consists of two main stages: a thermal and a chemical treatment.

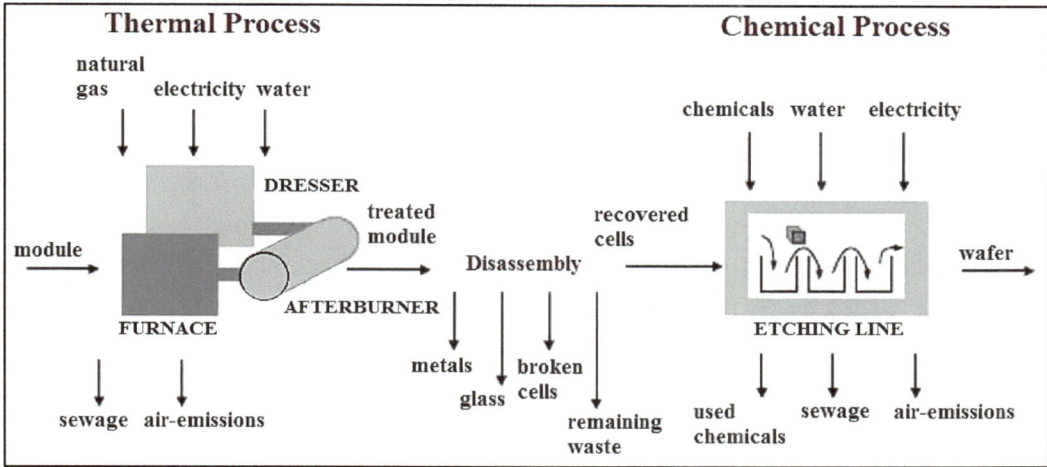

Figure 2: Recycling process of crystalline modules developed by Deutsche Solar.

In Fig. **3**, the two paths, in which the recycling process of Deutsche Solar branches, are shown: the main path which aims to recover intact wafer and the secondary path where broken wafers and solar cell breakage are exploited in order to obtain silicon to be used as secondary raw material (Si block manufacture). In this way the process objectives are twofold: on one side the undamaged wafers can be reused by solar cells producers, and on the other side it is possible to recovery silicon as raw material by damaged wafer and solar cells (Bombach *et al.*, 2005).

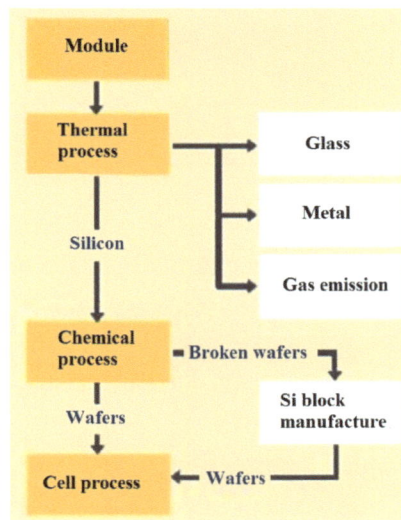

Figure 3: Recovery of intact wafers for reuse and recovery of broken wafers for ingot growth.

The thermal phase is carried out by mean of a small scale furnace and it is necessary and functional to the following steps of disassembly and separation of the various materials of the modules.

It should be noted that a recently developed automated separation system ensures very high separation efficiencies. As shown in Fig. **4**, PV modules are introduced to the thermal treatment system (characterized by higher energy efficiency than the furnace of the pilot plant) by an automated loading system (feeder). During this step, the decomposition of the organic materials, which serves to bind the different layers of the panel, occurs.

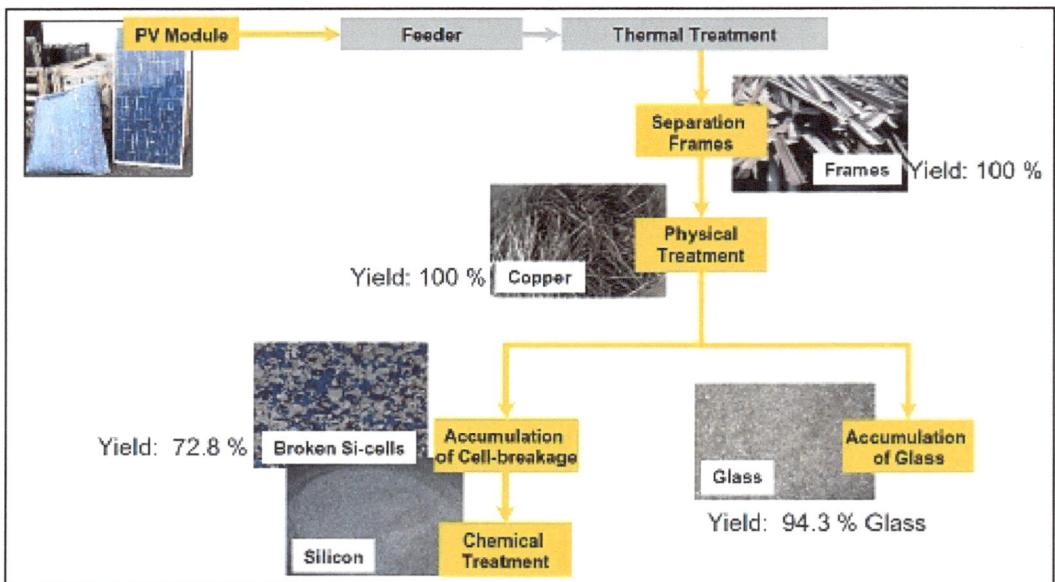

Figure 4: Automated recycling-process – Source: SolarWorld.

The thermal step is followed by the mechanical separation of the aluminium frame segments and by the electromagnetic separation of the copper strings from remaining materials. Due to their dimensions, these components can be completely recovered with an efficiency of 100%.

Afterwards, a set of mechanical treatments, like for example crushing, screening and gravimetric separation, allows to remove undesirable/pollutant materials and to separate glass from fragments of broken solar cells. Finally, a chemical treatment (etching), aiming at recovering silicon, is carried out on the glass fragments (Wambach *et al.*, 2009).

Also the damaged solar cells, after having been manually separated from other materials, are subjected to a chemical treatment (etching). This treatment allows separating and recovering the solar grade silicon from other materials that have been used for the coating of the wafers (in other words, the electrical contacts of the solar cells).

In both cases, the recovered silicon will be used as secondary raw material in the process of silicon ingot growth.

Through this system, it is expected that globally about the 95.7% of the mass of a PV module can be recycled.

Moreover, the 94.3% in weight of the glass fraction (with a purity of 99.99975%) and the 73% in weight of silicon fraction, corresponding to solar cells, can be recovered. Of this amount, the 59% can be obtained in a purity of 99.9999% and the remaining 41% in a purity of 99.995%.

The remaining 5.7% of the glass fraction is lost in the mixed material fraction, while the solar cells fraction, which cannot be recycled, is about 27%.

After all, only the 4.3% in weight of the module, corresponding to the sum of the finest fraction and of the amount of undesirable/pollutant materials, is not recovered.

RESOLVED Project

The RESOLVED project consists of a recycling strategy developed for CIS and CdTe PV modules, mainly based on the use of wet-mechanical treatment (Resolved, 2008). At present, the recycling of thin film PV modules, if they are recycled at all, is obtained by chemical processes.

The wet-mechanical treatment of end-of-life PV modules, developed by the RESOLVED project, can be considered as an innovative approach for PV recycling with a reduced use of chemicals (Fig. **5**).

The proposed approach consists in a "two-way" closed loop recycling strategy: the process of the loop described on the left is used for the recycling of intact end-of-life modules (without protection glass breakage) and production scraps (it is a sort of sub-module formed by a glass layer coated with a set of metal layers, containing the photoactive layer, placed one upon another), while the process of the loop described on the right can be used for the recycling of both undamaged and damaged modules.

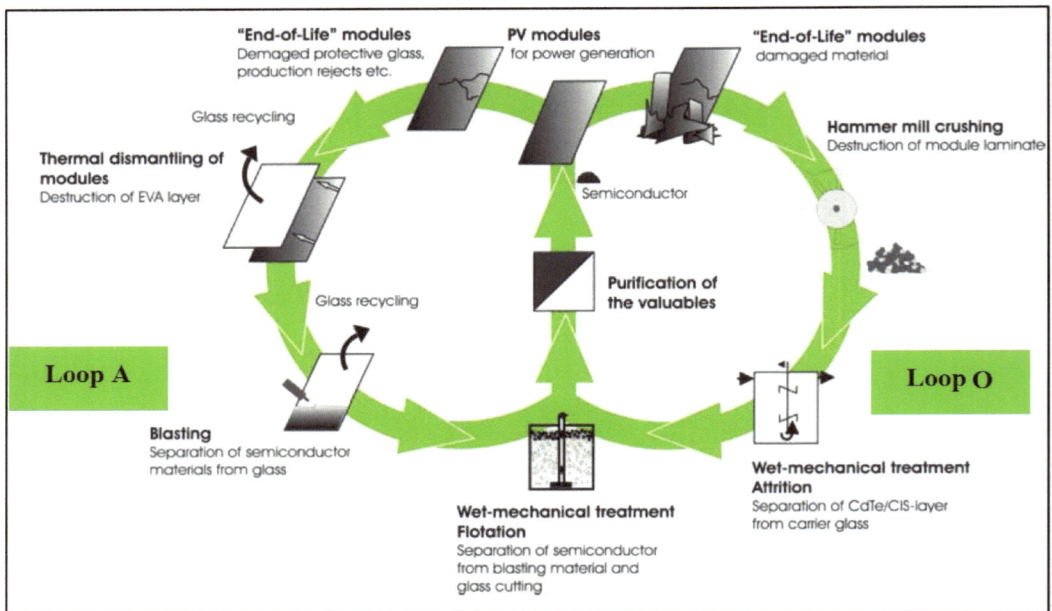

Figure 5: The "two-way" closed loop recycling strategy for thin film modules developed by RESOLVED project.

Recycling of Undamaged End-of-Life PV Modules (Counter-Clockwise Loop A, Fig. 5)

A1-Thermal treatment (thermal dismantling): The modules undergo a thermal pre-treatment in order to ensure an easy disassembly. During this step, which brings panels to temperatures within a range of 450-500 °C, the encapsulant layer of EVA is destroyed, in order to allow the separation of the module in two single glass plates. It should be noted that in the case of CdTe PV modules, the photoactive coating layer is deposited on the front glass coverage, while in the case of CIS PV modules the semiconductor layer is deposited on the back glass.

At this point, the glass plate coated with photoactive materials and with the metal layers, which form the electrical contacts, is separated from the other glass plate and sent to the next step, while the glass plate without coatings is decontaminated and sent to conventional glass recycling.

Obviously, in the case of sub-modules, the thermal treatment is not required because they are not encapsulated with the layer of EVA.

A2-Recovery of valuable materials (vacuum blasting): The glass plate coated with the semiconductor layer and the other metal layers is sent to a vacuum blasting process.

The power of the jet of vacuum blasting is much lower than the power of the jet obtained with the more common pressure system, because abrasive arrives onto the surface to be treated using vacuum instead of air pressure. Thus, thanks to a more gentle approach, this technology is more suitable than sand blasting for the treatment of glass surfaces.

Through the action of special abrasive materials, the semiconductor film and the other metal layers are removed from the glass surface, which, after having removed the last residual impurities, can be sent to conventional glass recycling.

Afterwards, the blasting fines, containing the semiconductor, as well as the blasting abrasives and glass fines, are collected by an industrial vacuum cleaner and treated by mean of wet-mechanical and physical-chemical processes, like for example flotation, in order to enrich valuable materials (CdTe or CIS) in a sort of pre-concentrate.

Recycling of Undamaged and Damaged End-of-Life PV Modules (Clockwise Loop O, Fig. 5)

This strategy, differently form the first one, is suitable to both undamaged and damaged PV modules. In fact, the treatment of small pieces of broken modules would be nearly impossible, using thermal and vacuum blasting processes. The main steps of the clockwise loop of Fig. **5** are described below.

O1-Pre-treatment by crushing: Firstly, both undamaged and damaged PV modules are crushed in a mill and reduced in particles of small size. This process leads to the destruction of the module structure from which the semiconductor film can be removed with a subsequent wet-mechanical treatment. The crushed materials consist of large fragments of EVA foils (that can be removed) and glass particles of different sizes with a semiconductor layer on the surface.

O2-Recovery of valuable materials by wet-mechanical treatment (attrition): The crushed materials, previously obtained, are subjected to a wet-mechanical treatment based on the use of shear and friction forces (attrition). During this step, the semiconductor layer is removed from the glass substrate using a high shear mixing system equipped with rotating blades (batch mixer).

The separation of the photoactive material requires a certain application of energy in the form of shear and friction forces.

After this treatment, photoactive materials with the finest particles of glass are removed from the glass substrate, while the larger particles of glass, at this stage no longer covered with the semiconductor film, can be sent to recycling.

The main benefit of this step is that there is no use of chemicals but only of water.

Flotation and Purification

The outputs coming from the two recycling loops described above (a mixture of fine glass particles, photoactive materials, other metals in general and, in the case of vacuum blasting, also abrasive materials) are treated using a flotation process, in order to recover the valuable semiconductors.

1-Enrichment of the valuable materials by flotation: Flotation is a wet-mechanical and physical-chemical process mainly used by mining industry to concentrate ores.

In this case, flotation is used to separate the semiconductor material from the rest of the mixture (coming from one of the two recycling loops previously described), and then to get a pre-concentrate for the final purification.

Flotation is carried out activating the surface of the semiconductor (which consequently becomes hydrophobic) so that it can be adsorbed by air bubbles, injected in the bottom of the flotation cells, and ascend to the surface. The foam, containing the valuable photoactive materials, gathering at the surface of the flotation cell, is then collected, while the fine glass particles deposit themselves on the bottom of the cell together with the tailings.

2-Purification: Since semiconductors used in thin film photovoltaic applications have to be of very high purity (99.999% or 5 N), the product coming from flotation has to be treated with a leaching process, which consists in the use of an acid solution to recover the photoactive materials.

Valuable materials are therefore recovered through precipitation, so that they can be subsequently reintroduced, as secondary raw materials, in the traditional production cycles of semiconductor materials for photovoltaic use.

COMPARISON BETWEEN "HIGH VALUE RECYCLING" SOLUTIONS AND "LOW VALUE RECYCLING" APPROACHES

All the approaches described above are examples of "high value recycling" strategies as they allow valorising valuable materials and components that can be found in the end-of-life PV modules. It is, therefore, appropriate to distinguish these "noble" stategies from other possible "low value" end-of-life management strategies, which, although easier to implement (therefore less expensive), are less attractive from an environmental point of view. As a matter of fact, "low value" recycling processes are deemed only to the recovery and recycling of some fractions, like for example the glass and the aluminium of the frame, without considering the valuable materials like silicon, indium and tellurium which are more difficult to recover.

These "low value" recycling strategies could compete only in some cases with the first ones. In the case of crystalline silicon PV modules, a possible low value recycling solution, which would be alternative to the process developed by Deutsche Solar, could consist in a pre-treatment of the panels in a municipal wastes incineration plant and in a subsequent disposal of the residues at a landfill for inert wastes (Müller *et al.*, 2005).

In fact, as wastes coming from end-of-life PV modules contain a very high mass percentage of organic materials, they cannot be disposed of in landfills without pre-treatment, at least according to European rules and procedures for the acceptance of waste in landfill sites. It should be noted that in the considered scenario only the aluminium of the frame would be recycled because the frame is likely to be recovered due to its economic value.

Therefore, if comparing this incineration approach with Deutsche Solar's recycling process, it is possible to note that a municipal waste incinerator usually is a large scale plant, unlikely to the furnace used by the PV module producer. Consequently the first one has much lower specific energy consumption than the second one, and this represents a benefit. However, it is also possible to understand that the incineration scenario has a relevant drawback, because it doesn't allow to benefit from the environmental advantages coming from the recovery of the wafers or more in general of the silicon, that would be inevitably lost in landfill.

DESIGN FOR RECYCLING AND PHOTOVOLTAIC

The "design for recycling" is a general design philosophy which aims to design products in such a way that it is relatively easy to disassemble a product, at the end of its useful life, into separate components that can be reused and recycled. However, with respect to solar panels, the "design for recycling" concept appears to be in conflict with the requirement of endurance that PV modules have to ensure, as they operate outdoors and must to be able to withstand to the weather (rain, snow, hail, moisture, *etc.*) to which they are constantly exposed for at least 25-30 years of useful life. On the other side it is evident that, if during the design phase of a PV module it could be possible to take into account, the future needs of disassembly, the end-of-life management would benefit from an economic and environmental point of view (in fact the environmental impacts of the thermal treatment would be avoided).

Therefore, on the basis of the aforementioned considerations, the application of "design for recycling" principles represents a real challenge for solar modules designers, as they have to solve a complex problem of trade-off between two conflicting needs: in fact on one hand a very reliable product, which has to be able to withstand to the weather for many years of useful life, is required, while on the other hand a product, that is relatively easy to disassemble at end of life, is desired.

CONCLUSIONS

Photovoltaic energy production has to be related with environmental benefits also at the end-of-life of the PV panels.

In fact, the PV panels contain valuable materials that could be recovered or recycled in view of a possible further use in the production of new PV modules or of other products.

Also silicon solar cells, if undamaged recovered, can be reused to make new recycled PV modules, ensuring a considerable saving in terms of energy and of raw materials (silicon). In the case of cadmium telluride PV modules, the recovery and the recycling of cadmium, a highly toxic material, reusable in the production of new modules, could reduce the environmental concerns due to the disposal of these modules in landfills.

In this way, it could be possible to move from the concept of waste to the concept of secondary raw material, which can be reused in a new production cycle (recycling), obtaining consequently benefits in terms of both preservation of natural resources of the Earth and reduction of energy demand.

ACKNOWLEDGEMENT

Declared none.

CONFLICT OF INTEREST

The author(s) confirm that this chapter content has no conflict of interest.

REFERENCES

Bianchi, M., Gambarotta, A., Peretto, A. (2004). *Impatto ambientale dei sistemi energetici, Vol 1.* Nuova Edizione, Pitagora Editrice, Bologna.

Bombach, E., Müller, A., Wambach, K., Röver, I. (2005). Recycling of solar cells and modules-recent improvements, In: *Final Proceedings of 20th EU PVSEC*, Barcelona, Spain.

First Solar (2010). *First Solar module collection and recycling program.* Available from: http://www.firstsolar.com/~/media/WWW/Files/Downloads/PDF/Document-Library/Sustainable-Development/Brochure_CollectionRecyclingProgram.ashx?la=en&fstk=/DocLibrary/Download/Brochure_CollectionRecyclingProgram_NA.pdf [Accessed 17th February 2012].

Fthenakis, V.M., Eberspacher, C. (1997). Disposal and recycling of end-of-life PV modules, In: *Final Proceedings of 26th PVSC*, Anaheim, CA, 1067-1072.

Gabrielli, F. (2010). *La sfida del riciclaggio nel settore fotovoltaico*. Nextville Energie Rinnovabili ed Efficienza Energetica. Available from: http://www.nextville.it/scenari/14 [Accessed 17th February 2012]

Müller, A., Wambach, K., Alsema, E.A. (2005). *Life cycle analysis of a solar module recycling process*, In: Final Proceedings of 20th EU PVSEC, Barcelona, Spain, 3211-3213.

Resolved (2008) *Recovery of solar valuable materials, enrichment and decontamination.* Available from: http://www.resolved.bam.de/eng_publications.htm [Accessed 17th February 2012].

Sander, K., Schilling, S., Reinschmidt, J., Wambach, K., Schlenker, S., Müller, A., Springer, J., Fouquet, D., Jelitte, A., Stryi-Hipp, G., Chrometzka, T. (2007). *Study on the development of a take back and recovery system for photovoltaic products*, Oekopol Gmbh, BMU Project Report (co-financed by EPIA and BSW solar), 03MAP092, Hamburg. Available from: http://www.PVcycle.org/index.php?id=45 [Accessed 17th February 2012].

Wambach, K., Schlenker, S., Konrad, B., Müller, A., von Ramin-Marro, D., Clyncke, J., Gomez, V., Hartleitner, B., Rommel, W. (2009). PV Cycle – The voluntary take back system and industrial recycling of PV modules, In: *Final Proceedings of 24th EU PVSEC*, Hamburg, Germany, 4417-4421.

Send Orders of Reprints at reprints@benthamscience.org
Separating Pro-Environment, 2012, 109-122

CHAPTER 7

The Control of Separation Processes in Mechanical Recycling of Waste Refrigerators by Partition Function

F. La Marca[*]

Department of Chemical Engineering Materials & Environment, University of Rome "La Sapienza", Rome, Italy

Abstract: The recycling of e-waste in general allows the recovery of valuable materials, which can be reused as secondary raw materials. However this implies, to adopt reliable treatment processes to achieve specific standard characteristics. Actually, the market standards are very strict and strongly affect the economic value of recovered materials.

In particular, waste refrigerators recycling allows the recovery of different plastic materials and metallic fractions (ferrous and non ferrous), whose economic value has considerably increased in the last years. Only in Italy, in 2010, approximately 66,000 tons of waste refrigerators were collected for recycling. An efficient recycling treatment could assure about 85% of recovery rate.

In this paper, a model to control and characterize the materials recovered from mechanical recycling of carcasses dismantled from waste refrigerators is proposed. The treatment flow sheet consists of various separation and comminution operations. A mathematical model was implemented to determine a partition function, utilizing data obtained from the treatment of 100 waste refrigerator carcasses, carrying out mass balance of recovered materials. The partition function was used to determine control parameters and assess the quality of the recovered products, and, therefore, of the applied technologies. The results have shown that for waste refrigerator carcasses it has been possible to reach for ferrous metals a recovery rate of 97.5% with purity of more than 98%. Other metals also reached a good quality (about 87% for non-ferrous metals, 92% for mixed recyclable plastics), but with lower recovery rates.

Keywords: Material control and characterization, Mechanical recycling, Waste refrigerators.

INTRODUCTION

The implementation of policies aiming at low-carbon economy and extended producer responsibility (EPR) principle in the sector of electrical and electronic

*Address correspondence to F. La Marca: Department of Chemical Engineering Materials & Environment, University of Rome "La Sapienza", via Eudossiana 18, 00184 Rome, Italy; Tel: +390644585615; Fax: +390644585618; Email: floriana.lamarca@uniroma1.it

equipment (EEE) increased the waste stream to recycling facilities (European Commission, 2002). According to Eurostat (Eurostat, 2011), in 2008 the amount of large household appliances waste collected in UE was about 1.8 million of tons, with a recovery and reuse rate of about 69%.

The large household appliances may include air conditioners, dishwashers, refrigerators, washing machines, *etc.* The interest in recovery of materials from waste of large household appliances is growing, especially in the last decades (Penev and de Ron, 1994; Laner and Rechberger, 2007; Deng *et al.*, 2008 and Ruan and Xu, 2011). However, the recycling treatment of each typology of appliances requires dealing with specific issues.

This paper is focused on the recycling treatment for waste refrigerators.

Waste refrigerators contain abundant recyclable resources, such as metals (steel, copper, aluminum) and non-metals (ABS, PVC, glass, polyurethane foam). Nevertheless, waste refrigerators comprise also hazardous CFCs, stored in the compressor, as refrigerant gases, and in polyurethane (PUR) foam, as blowing agent, which may destroy the stratospheric ozone layer when they are released into the atmosphere (The Ozone Hole, 2006). Additionally, 1 kg of CFCs has a global warming potential of approximately 11,000 kg carbon dioxide. For these reasons, waste refrigerator management has to obey to severe national and international standardized procedures.

This study was developed in collaboration with a recycling plant located in the surrounding area of Rome, Italy. The plant is privately operated and accomplishes all standard requirements, as defined in recent Italian national legislation (DLgs 151/2005). The recycling treatment for waste refrigerators is certified in relation to the storage and processing of WEEE containing CFC, HFC, HCFC and HC, according to ISO14001(International Organization for Standardization, 2004).

The objective of the recycling plant is the recovery of ferrous and non ferrous metals, of plastics and polyurethane foam. A proper management of pollutants through physical and thermal-mechanical processes is also carried out. The complete scheme for treatment of waste refrigerators can be divided up to the three phases below described.

1. *Safety*: the first step of the process is to remove valuable components, such as electric cables, plastic and glass food cases, compressors, metallic and wood fractions. Such operations are completed manually. Then the refrigerant gases are drained from the cooling circuit and collected in cylinders in order to be sent to the specific treatment for disposal.

2. *Material recovery by mechanical recycling*: the refrigerator carcasses are then fed to the recycling plant, consisting of comminution and separation processes aimed to produce secondary raw materials, whose characteristics have to comply with market standards. Comminution processes are carried out in enclosed and depressurized environment to remove the CFCs liberated from polyurethane foams.

3. *CFC recovery*: in accordance with statutory prescriptions, CFCs extraction is monitored and recorded during all processing. The CFCs stored in cylinders and sucked up during carcasses comminution are conveyed in a cryo-condensation unit with activated carbons.

MATERIALS AND METHODS

After removing valuable components and collecting the refrigerants, the material recovery by mechanical recycling of refrigerator carcasses is carried out by an automated process. The flow-sheet of the applied technology is shown in Fig. **1**: each operation is numbered from (0) to (7), while products are associated to a code, which refers to origin and derived processes, in blue for intermediate products, in green for final products and in red for waste products.

It originates from manual dismantling (0) and includes two stages of comminution in enclosed and depressurized environment (1) and (5), magnetic separation (2), eddy current separation (3), manual separation (4), vibrating table separation (6), and cycloning (7).

The recovered products are ferrous metals (*p2Fe*), copper (*p4Cu*), other non-ferrous metals (*p4NFe*) constituted mainly by aluminium, mixed recyclable plastics (*p6PLA*) and polyurerhane (*p7PU*). Furthermore a waste product (*p7W*) is generated.

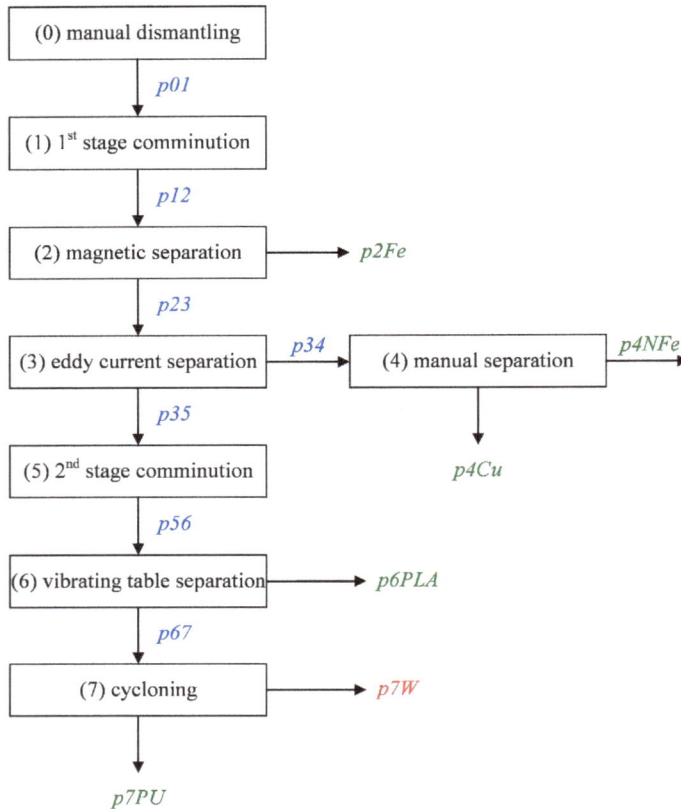

Figure 1: Flow-sheet of the applied technology.

The comminution in enclosed and depressurized environment is used to liberate the different materials composing the waste refrigerator carcass, avoiding the dispersion of the refrigerant contained in the polyurethane foam. The refrigerants are recovered and conveyed in a cryo-condensation unit with activated carbons. The comminution process is carried out by shear and impulsive force in order to crush refrigerator carcasses as the materials have thin shape and great two-dimensional scale. In the first stage of comminution process, a twin shafts shredder, followed by a four shafts shredder, both equipped with blades, produce particles 20-30 mm in size. A perforated plate controls particle size of the output product. In this stage, metal faceplates, plastic inner film, polyurethane foam insulating layer, and metal wires are liberated. The output of the comminution process is transported to the magnetic separation, for sorting steel or ferrous metals, followed by eddy current separation, for recovering non-ferrous metals (aluminium, copper, *etc.*).

The magnetic separator consists in an electromagnet facing the belt conveyor. Ferrous particles are attracted by the electromagnet if the magnetic force is greater than the opposing component of the gravitational force in that specific direction. The ferrous particles are then detached from the main stream and collected in a separate tank (Fig. **2**), while non-ferrous particles (mainly non ferrous metals, plastic and polyurethane) remain on the belt conveyor and proceed to the eddy current separator.

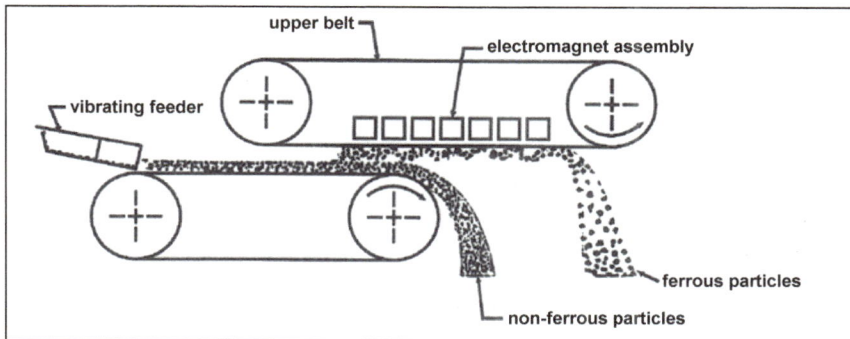

Figure 2: Schematic diagram of magnetic separation.

Eddy currents are induced in nonferrous metal when plunged in a variable magnetic field. Such currents generate a secondary magnetic field in the particles that interacts with the variable magnetic field, determining a repulsive force on non-ferrous metals (mainly aluminum and copper) and separating them from non-metallic particles (mainly plastic and polyurethane) (Fig. **3**).

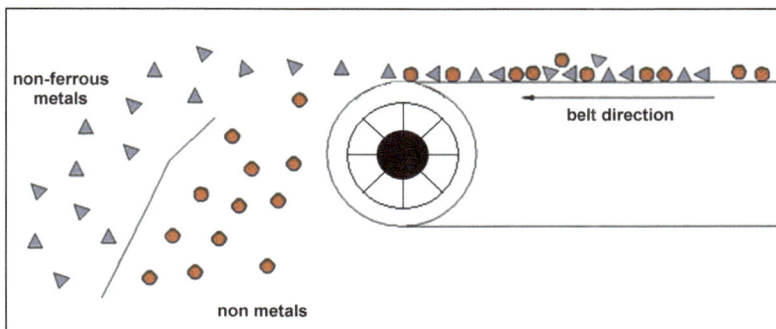

Figure 3: Schematic diagram of eddy current separation.

A second stage of comminution process is then carried out by a knife mill to reduce plastic and polyurethane in particles 5-10 mm in size. A vibrating table is employed to separate polyurethane foam from other plastic particles, based on their difference in density. The particles to be separated are fed onto the separation table, under which an adjustable air flow is generated. The combined effects of table vibration and air flow from below reduce the friction between the particles, so that particles behave like a fluid. Due to the vibration, the heaviest materials tend to move upwards on the slightly sloping table, while the lighter particles flow downward carried by the ascending air stream. Finally, separation by cycloning is designed for removing very fine particles from the polyurethane product. Cyclone separators utilize centrifugal forces and two opposing air flows caused by spinning motion to separate particles lower than 125 μm in size from the final polyurethane product.

THE PARTITION FUNCTION

In recycling operations, a separation process is aimed to identify a target product from a feed flow on the basis of materials properties and composition.

In binary separation processes, the flow-sheet of products transformation can be represented as in Fig. **4**, considering the balance of total mass.

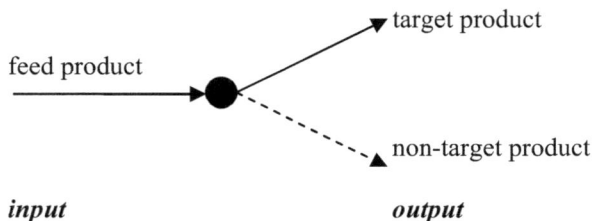

Figure 4: Flow-sheet of products transformation in binary separation process.

Generally, the mass balance of products involved in the separation processes is expressed by the following relation:

input − output = accumulation   **(1)**

In case of steady state in a continuous system, the input material flows to and from a single separation operation can be considered constant, thus the mass balance of products involved in the separation processes is simplified as follows:

input − output = 0 (2)

To provide a modelling of separation processes, the material flows can be represented by a pattern vector related to its composition, composed by *n*-elements representing the weight percentage of the *n* material typologies.

For each *i-th* material typology, the mass balance of the product of the separation process is represented in the Fig. **5**, being:

$p_{t(i)}$ mass of particles of *i-th* material typology in the target product;

$p_{nt(i)}$ mass of particles of *i-th* material typology in the non-target product;

$p_{0(i)}$ mass of particles of *i-th* material typology in the feed product.

The analysis of the results of a separation process is based on the evaluation of a *partition function*, representative of the utilized device. From the separation function, furthermore, a set of parameters can be calculated to estimate the efficiency of the applied technologies and to control the quality of the recoverable products.

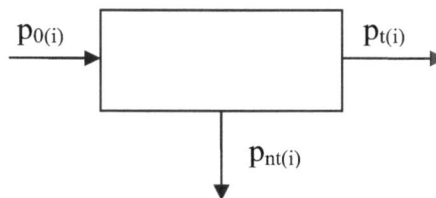

Figure 5: Flow-sheet of material transformation in binary separation process.

For each considered material typology, considering the operative conditions of a separation device, the partition function is defined as:

$$\alpha(i) = f[\delta(i)] \tag{3}$$

being:

    $\alpha(i)$ frequency with which a particle of composition $\delta(i)$ is selected in the target product;

    $\delta(i)$ *i-th* material typology.

In order to define the partition function, it is necessary to take a sample of the products obtained in the separation process. Each sample, representative and proportional to the product under investigation, is analyzed to determine the material composition, in relation to the n considered typologies, measuring the values of $p_{t(i)}$, $p_{nt(i)}$ and $p_{0(i)}$.

For each separation device, the partition function is analyzed by the following mathematical model:

$$\alpha_{(i)} = \frac{p_{t(i)}}{p_{0(i)}} = \frac{p_{t(i)}}{p_{t(i)} + p_{nt(i)}} \tag{4}$$

An ideal partition function, $\alpha_{id(i)}$ can be defined, referring to the perfect separation, which values are equal to 1.00, for the *i-th* material typology expected in the target product (*e.g.* ferrous metals in the magnetic separation) and are equal to 0.00 for the *i-th* material typology expected in the non-target product. The partition function for each separation process is determined, starting from the measured mass of the products, according to the proposed mathematical model. Once implemented the partition function, the defined model can predict the output products composition in relation to different composition of the input feed, considering the same applied technology. To quantify the separation efficiency in a separation process, the following quantities are considered, calculated from the partition function:

-     The grade of target material in the target product, equal to the ratio between the weight of recovered target material and the weight of total target product, expressed as a percentage;

$$g_{(i)t} = \frac{P_{t(i)}}{\sum P_{t(i)}} = \frac{P_{t(i)}}{P_t} \qquad (5)$$

- The recovery of target material, equal to the ratio between the weight of the recovered target material in the target product and the weight of the target material in the feed, expressed as a percentage;

$$r_{(i)t} = \frac{P_{t(i)}}{P_{0(i)}} \qquad (6)$$

The grade provides an indication of the target material and contaminant concentration in the target product, while the recovery allows for comparison between the amount of target material in the target product and in the feed product.

CASE STUDY

The case study involved 100 waste domestic refrigerators, processed by a recycling facility, located in the surrounding of the city of Rome, Italy. The number of refrigerators is adequate to assure a representative average composition. Particularly, the automated process on refrigerator carcasses was investigated. The average composition of the refrigerator carcasses is reported in Table **1**.

Table 1: Mass balance of treated refrigerator carcasses, in percentage

Material Typology	%Weight
ferrous metals	52.0
non-ferrous metals (excluding copper)	7.0
copper	0.5
mixed plastics	14.0
polyurethane	10.0
other	16.5
TOTAL	100.0

The total weight of the considered waste refrigerator carcasses was about 4,564.6 kg, consisting of ferrous metals, nonferrous metals, plastics, PUR foam. The weight of CFCs can be negligible, being lower than 0.1%. For each separation process, the composition of feed, target and non-target products was determined by laboratory analysis of characterization. Such analyses fall within the ordinary activities to control industrial results and product quality, notably for secondary raw materials. In the following, for each separation process, the analyses for the characterization of the products are listed:

- *Magnetic Separation (MagS)*: by a permanent magnet, to evaluate the ferromagnetic metallic fraction content;

- *Eddy Current Separation (ECS)*: by electrostatic separation at lab scale, to evaluate the non-ferrous metallic fraction content;

- *Manual Separation (ManS)*: by visual examination, to evaluate the copper fraction content;

- *Vibrating Table Separation (VTS)*: by picnometry and IR-spectrometry, to evaluate the content and typologies of the various plastic material fractions (mixed plastics and polyurethane foam);

- *Cycloning (Cyc)*: as for vibrating table separation, by picnometry and IR-spectrometry, to evaluate the content and typologies of the various plastic material fractions (mixed plastics and polyurethane foam).

Table 2: Mass data and partition function data related to the magnetic separation process

Material Typology	*Products*			*Partition Function*
	Target (%)	*Non-Target (%)*	*Feed (%)*	*Real (/)*
$\delta_{(i)}$	$p_{t(i)}$	$p_{nt(i)}$	$p_{0(i)}$	$\alpha_{(i)}$
Fe-metals	50.72	1.28	52.00	0.98
non-Fe metals	0.15	6.85	7.00	0.02
Cu	0.14	0.36	0.50	0.29
rec-PLA	0.10	13.90	14.00	0.01
PU	0.07	9.93	10.00	0.01
other	0.49	16.01	16.50	0.03
Total	51.68	48.32	100.00	

From the measured data on the composition of feed, target and non-target products for each considered material typology, the partition function was computed, according to equation (4). In Table **2** an example of measured mass and partition function data, related to the magnetic separation process are given, while in Table **3** only the partition function data related to the other separation processes are reported.

Table 3: Partition function data related to the other separation process

	ECS	*ManS*	*VTS*	*Cyc*
Material Typology	*Real (/)*	*Real (/)*	*Real (/)*	*Real (/)*
$\delta_{(i)}$	$\alpha_{(i)}$	$\alpha_{(i)}$	$\alpha_{(i)}$	$\alpha_{(i)}$
Fe-metals	0.00	0.00	0.55	0.84
non-Fe metals	0.71	0.00	0.05	0.92
Cu	0.76	1.00	0.08	0.74
rec-PLA	0.02	0.00	0.86	0.78
PU	0.00	0.00	0.01	0.86
other	0.02	0.00	0.01	0.45

RESULTS AND DISCUSSION

The grade and recovery values obtained in the target products for each separation process (Table **4**) were calculated starting from the data measured by lab analyses, used also for the calculation of the partition function. The magnetic separation shows satisfactory results in terms both of grade and of recovery on the metallic fraction (*p2Fe*). The obtained ferrous product is characterized by very good quality (grade > 98%), with high recovery, and then high efficiency. The efficiency of eddy current separation is quite satisfactory too, showing adequate results in the target product (*p34*), being grade equal to 87.74% and recovery equal to 71.25%.

Manual separation is aimed, in particular, to recover copper, because of its current high quotation. The separation by vibrating table is very efficient in sorting mixed recyclable plastics (*p6PLA*, grade equal to 92.00%), while the quality of polyurethane product (*p67*) is significantly lower. As a matter of fact, the polyurethane product is highly contaminated by waste fractions. A slightly improvement of its quality is achieved after cycloning, which allows to increase the grade from 32.90% to 43.88%.

Table 4: Grades and recoveries of target products for each separation process (refer to Fig. **1** for product codes)

	Separation Process	Code	%Grade	%Recovery
(2)	magnetic	p2Fe	98.15	97.54
(3)	eddy current	p34	87.74	71.25
(4)	manual (non-Fe metals)	p4NFe	87.15	100.00
(4)	manual (Cu)	p4Cu	100.00	100.00
(6)	vibrating table (rec-PLA)	p6PLA	92.00	85.79
(6)	vibrating table (PU)	p67	32.90	98.79
(7)	cycloning (final PU)	p7PU	43.88	86.33

To define the performance of the complete recycling treatment, the recovery values of target materials were evaluated on the process as a whole. Table **5** shows the recovery values for each target product derived from the whole recycling process, together with the grade values.

Table 5: Grades and recoveries of target products for the whole recycling process (refer to Fig. **1** for product codes)

Products	Code	%Grade	%Recovery
ferrous metals	p2Fe	98.15	97.54
non-Fe metals	p4NFe	87.15	69.50
copper	p4Cu	100.00	53.78
mixed recyclable plastics	p6PLA	92.00	83.22
polyurethane	p67	32.90	98.07
de-dusted polyurethane	p7PU	43.88	84.67

In the ferrous metal product (*p2Fe*), the contamination is lower than 2%, as required by market standards for ferrous scraps. The overall recovery is very high (97.54%), due to the high efficiency of magnetic separation. The 69.50% recovery of non ferrous product (*p4NFe*) and 53.78% of copper (*p4Cu*) are due to the loss of smallest particles and/or wire fragments, which are difficult to handle. The value of total recovery referred to mixed recyclable plastics (*p6PLA*) is quite high, but such product is composed by a mixture of different plastics, so its economic value is not very high. The polyurethane is quite completely recovered in the final products *p67* and *p7PU* before and after cycloning, respectively, but its quality is very poor. For this reason, such products are still considered waste materials.

CONCLUSIONS

This study has investigated a recycling automated process for the recovery of materials which constitute the carcasses of waste refrigerator. The process includes comminution, in enclosed and depressurized environment, and separation processes. The analysis was focused on the efficiency of the separation process, starting from the physico-chemical analysis of the inputs and outputs of the process and on the modelling of a partition function. The same data used to calculate the partition function allowed the definition of the parameters related to the results of the different separation processes (grade and recovery).

Analysing the results, as predictable, the magnetic separation seems to be the most efficient technology, but, in general, all the separation devices achieve good results, in terms of quality of the target products, measured by the grade values. For mixed recyclable plastics, the different recycling/recovery options need to be evaluated, primarily focusing on their environmental compliance, considering industrial processing for non homogeneous plastics or energy recovery.

Only the polyurethane product quality is not satisfactory, so it should not be materially recycled. For this reason, it is considered a waste product even after treatment.

The proposed approach to separation processes control seems to be feasible and cost-effective, in term of both data availability and simplicity in industrial application. As a matter of fact, the required analyses for the modelling of the partition function are ordinarily carried out off-line at the recycling plant on products recovered as secondary raw material, to determine the compliance to market standards. Furthermore, the comparison between real and ideal partition function allows the monitoring of separation devices and their operative conditions.

ACKNOWLEDGEMENT

Declared none.

CONFLICT OF INTEREST

The author confirms that this chapter content has no conflict of interest.

REFERENCES

Deng, J., Wen, X., Zhao, Y. (2008). Evaluating the treatment of E-waste – a case study of discarded refrigerators, *Journal of China University of Mining & Technology,* 18,. 454-458.

DLgs 151/2005 (2005) *Normativa specifica contiene gli obblighi di gestione e finanziamento, in capo ai produttori di AEE, delle operazioni di ritiro, trasporto e gestione dei RAEE domestici e anche di raccolta dei RAEE professionali,* transposition of Directives 2002/95/CE 2002/96/CE e 2003/108/CE).

European Commission (2002) *Directive 2002/96/EC of the European Parliament and of the Council on waste electrical and electronic equipment* (WEEE).

Eurostat. (2011). *Waste electrical and electronic equipment (WEEE) collected and exported.* Available from: epp.eurostat.ec.europa.eu/portal/page/portal/waste/data/wastestreams/weee epp.eurostat.ec.europa.eu [Accessed 23rd September 2011].

International Organization for Standardization (2004) *ISO 14001:2004 Environmental management systems. Requirements with guidance for use.*

Laner, D., Rechberger, H. (2007). Treatment of cooling appliances: interrelations between environmental protection, resource conservation, and recovery rates. *Resource, Conservation and Recycling, 52*, 136-155.

Penev, K.D., de Ron, A.d.J. (1994). Development of disassembly line for refrigerators. *Industrial Engineering, 26*(11), 50–53.

Ruan, J., Xu, Z. (2011). Environmental friendly automated line for recovering the cabinet of waste refrigerator, *Waste Management, 31*(11), 2319-2326.

The Ozone Hole (2006). *Ozone Hole Consequences. Ozone Hole.* Available from: www.theozonehole.com/consequences.htm [Accessed 23rd September 2011].

Send Orders of Reprints at reprints@benthamscience.org
Separating Pro-Environment, 2012, 123-138

A Matrix Based Approach for Modelling of Treatment Processes for Contaminated Groundwater

V. Gente[1,*] and D. Lausdei[2]

[1]*Environmental Engineer, Italy and* [2]*ENVIRON Italy S.r.l., Rome, Italy*

Abstract: When dealing with contaminated sites, one of the main problems to be faced is the treatment of groundwater. In fact, in order to contain the contamination and to remediate the site, a common managing and operational option is to extract the groundwater and treat it at the surface prior to discharge. This remedial strategy is referred to as conventional pump-and-treat technology and its purpose is to remove or reduce contaminants concentrations so that, after the discharge of the treated water, the receiving waters meet the relevant quality objectives and provisions required by local or national regulations, directives or standards. Towards this aim, in the present research work, a matrix based approach has been adopted for modelling a treatment process for contaminated groundwater. In particular, it has been considered a modified Broadbent and Callcott breakage-matrix whose elements have been replaced by first-order kinetic equations that represent the depletion rate of contaminants' concentrations. The adopted approach allows determining the proper recirculation ratio in a groundwater treatment process for achieving the water quality objectives for discharge. Field tests have been conducted for the setting up of the mathematical model to be used for the revamping of an evaporating tower. The results of chemical analyses carried out on groundwater samples, collected in the input and the output flows of the evaporative tower, confirmed the effectiveness of the proposed approach.

Keywords: Contaminated sites, Groundwater treatment, Breakage-matrix.

INTRODUCTION

Site remediation options usually involve the treatment of both contaminated soil and groundwater, which can constitute the primary or secondary source of contamination for a given site.

When contamination concerns groundwater, the risk posed to the environment and to the population can increase due to the contaminants' transport from the

---

Address correspondence to V. Gente: Environmental Engineer, Italy; Tel.:+39069699532; Fax: +39069699532; E-mail: vincenzo.gente@ingpec.eu

source/release areas, such as abandoned landfills or leaking tanks, to down-gradient areas.

A common mean to contain and or remediate contaminated groundwater is to extract the water and treat it at the surface prior to discharge. This remedial strategy is referred to as conventional pump-and-treat (P&T) technology (U.S. EPA, 1990; U.S. EPA 2008).

P&T systems are frequently designed in order to hydraulically control the movement of contaminated groundwater and to prevent the expansion of the contamination zone (Fig. 1).

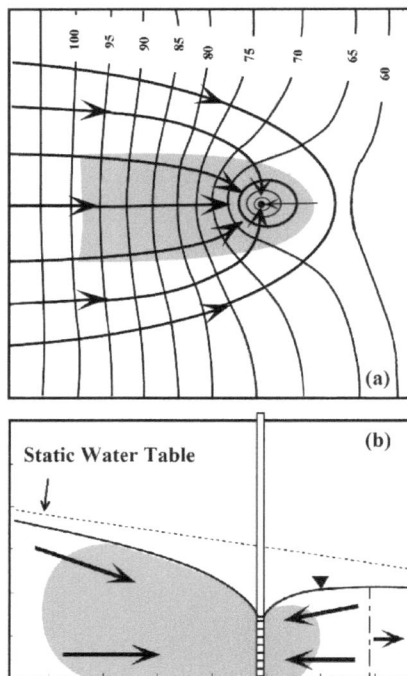

Figure 1: Examples of hydraulic containment in plan view (a) and cross section (b) using an extraction well (U.S. EPA, 1990 modified).

P&T systems present a large variety of configurations due to the fact that groundwater extraction and treatment processes are tailored to site specific conditions, to remediation goals and to local regulations or quality standards for discharge water objectives. Nevertheless, P&T systems include some main components that can be grouped in the following categories: pumping wells or drains; piping; treatment units, and receiving facility sewer systems or water bodies.

Pumping wells are intended to capture and remove contaminated groundwater, which is conveyed to the treatment units by mean of a piping system. In the treatment units, groundwater is physically, chemically, and/or biologically processed to reduce contaminants' concentrations to levels that are suitable for water discharge into facility sewer systems or water bodies. Examples of treatment technologies that can be used as stand-alone or as combined units are: air stripping, granular activated carbon (GAC), chemical/UV oxidation, aerobic biological reactors, chemical precipitation, ion exchange/adsorption, and electrochemical methods.

P&T performance is typically assessed by measuring hydraulic heads and gradients, ground-water flow directions and rates, pumping rates, quality of pumped and treated water, and contaminants' distributions in ground water and porous media (Cohen *et al.*, 1977; Javandel and Tsang, 1986. Endres *et al.*, 2007).

Samples of discharge water have to be collected and analysed periodically in order to evaluate that extracted groundwater is appropriately treated and that, after discharge, the receiving waters meet the relevant quality objectives and provisions of local or national regulations, directives or standards.

Due to slow transport of contaminants and interphase transfer, many P&T systems might operate for years to contain and clean up contaminated groundwater. Therefore, managing and operational data have to be periodically monitored to identify modifications that can occur to the site contamination and to the hydraulic conditions and that can affect the performances of P&T systems.

In fact, treatment units should be designed and managed considering the possibility of adapting the P&T systems to changing conditions over their whole operational life.

In the present paper, a research has been carried to model the functioning of a groundwater treatment process. In particular, the suggested model allows adjusting the performances of the considered treatment process modifying the recirculation ratio of treated groundwater before being discharged. The model can be therefore applied in case of existing treatment devices when changes occur in:

- Contamination characteristics (typology and\or concentration of contaminants);

- Site hydro-geological and hydraulic characteristics;

- Remediation objectives;

- Water quality standards for discharge;

- Or in case of new treatment devices when:

- The design is limited by the availability of space for the realization of the device itself (*e.g.* when the treatment unit has to be located inside existing premises);

- Achieving the water quality standards for discharge would require unfeasible dimensions of the treatment units.

For the modelling of the treatment process, an approach has been adopted based on the similarity of contaminants' removal or depletion with solid particle size-reduction processes. In particular, the breakage-matrix analysis proposed by Broadbent and Callcott, 1956, has been considered as reference for the mathematical model. Broadbent and Callcott analysed particle assemblies submitted to processes of size reduction and size classification, and described the particle size distribution by mean of vectors, and alterations to size distributions during breakage processes (crushing and grinding process) by mean of matrices.

The elements of the breakage matrices are calculated with a breakage function that depends on the characteristics of the crushing and grinding machines and on the nature of the particle assembly.

Likewise, in the case of treatment of contaminated groundwater, in this research work the contaminants' concentrations are described by mean of vectors, and modifications in contaminant concentrations during the treatment processes by mean of matrices multiplying the vectors. In this case, the elements of the matrix are replaced by depletion coefficients that depend on the contaminant and treatment process typologies. In particular, the adopted depletion coefficients have been calculated under the hypothesis that the removal of contaminants can be represented by kinetic equations of first order reaction.

The matrix approach has been applied for the revamping of an evaporative tower for the treatment of contaminated groundwater for three main contaminant typologies (iron, manganese and volatile organic compounds). The results of chemical analyses carried out on samples of groundwater, collected in the input and in the output flows of the evaporative tower, confirmed the effectiveness of the proposed approach.

MATERIALS AND METHODS

A schematic representation of a treatment system for contaminated groundwater is shown in Fig. **2**, where:

- **I, U, T, S** and **O** are pattern vectors representing, respectively, the overall inlet, the treatment unit inlet, the tower outlet, the recirculation charge and the overall outlet; the components of each vector, c_i, represent the concentrations (%) of the considered contaminant typology;

- q_i are the flow rates (l/min) of the water in the different circuits of the system;

- **R** is the recirculation pattern vector, whose components represent the recirculation percentage of each contaminant typology and it is consequently $r_1 = r_2 = \ldots = r_n$;

- **M** is the treatment matrix, whose components m_{ij} are determined on the basis of first-order kinetic equations.

In the Broadbent and Callcott model, the element b_{ij} of the breakage matrix **B** represents the proportion of a particle with dimension between a_{j-1} and a_j, with

$a_j < a_{j-1}$ before breakage, which falls between dimension a_{i-1} and a_i, with $a_i < a_{i-1}$ after breakage (i,j=1,..., n); as the size of a particle cannot be increased by breakage, $b_{ij}=0$ for i<j, consequently **B** is a lower triangular matrix.

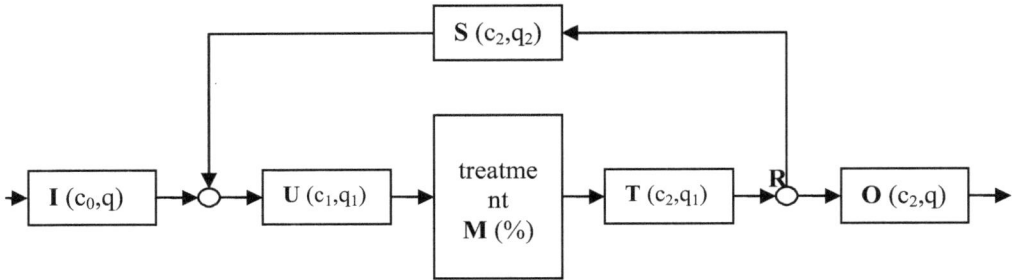

Figure 2: Schematic representation of system functioning.

Similarly, in the model of the groundwater treatment system, the element m_{ij} of the treatment matrix **M** represents the proportion of a contaminant p_j present in concentration values c_{j-1}-c_j before treatment which is depleted into contaminant p_i in concentration between c_{i-1}-c_i after treatment (i,j=1,..., n); under the hypothesis that the treatment would only reduce the concentration of each contaminant p_i without transforming it into other contaminants, it is $m_{ij}=0$ for i≠j, and consequently **M** is a diagonal matrix.

Therefore, considering a first order kinetic equation, $c = c_0 \cdot e^{-kt}$, in the matrix **M** the element m_{ij}, with i=j, will be equal to $c_i / c_{0i} = e^{-k_i t}$, where c_i and c_{0i} are respectively the concentrations of contaminant p_i at time t and t=0, and k_i is the kinetic coefficient for contaminant p_i. The element m_{ij} depends on the characteristics of contaminants and treatment process, through the coefficients k_i, and on the residence time t of the contaminated water in the treatment system.

In order to fully describe the treatment process it is necessary to border the pattern vectors with an extra row, and the treatment matrix **M** with an extra row and an extra column. The elements of these extra rows and columns are necessary to consider, along with the contaminants, the presence in the system of the water; they are for the pattern vectors and for the treatment matrix respectively the 100-complement and 1-complement of the sum of elements of the n above rows. Therefore a generic pattern vector **C** and the matrix **M** can be described as follows:

$$\mathbf{C} = \begin{vmatrix} c_1 \\ c_2 \\ \dots \\ c_n \\ \left(100 - \sum_n c_i\right) \end{vmatrix}, \ \mathbf{M} = \begin{vmatrix} m_1 & 0 & \dots & 0 \\ 0 & m_2 & \dots & 0 \\ \dots & \dots & \dots & \dots \\ 0 & \dots & m_n & 0 \\ 1-m_1 & \dots & 1-m_n & 1 \end{vmatrix}.$$

Having defined the elements of the above functioning scheme, it is possible to apply a mass balance to the different units showed in Fig. **2** and obtain the following array equations:

$$U = I + S \tag{1}$$

$$T = M \times U \tag{2}$$

$$O = (1 - R)\,T \tag{3}$$

$$S = R \times T \tag{4}$$

Once calculated the values of the coefficients k_i for the considered typologies of contaminants, it is then possible to evaluate, by iterative matrix calculations, the recirculation ratio **R** (defined as the portion of treated water to be recirculated, **S**, divided by the total amount of treated water, **T**), that allows obtaining an outlet complying with the water quality standards.

To set-up the matrix approach it is therefore necessary to determine the coefficients k_i for the considered contaminants. At this aim, in this research work, field tests were conducted on an evaporative tower, as described in the following paragraph.

Field Tests

The field tests were conducted on a P&T system in which the treatment unit is constituted by an evaporative tower. Installed in an industrial facility located within an agricultural area in Italy, the P&T system is design in order to create an hydraulic barrier to contain the contaminant inside the site through the extraction of groundwater from four barrier wells.

The evaporative tower is made up of a filling pack in PVC mounted in a fiberglass container. The contaminated water extracted from the pumping wells is sprayed through the nozzles onto the filling pack while a counter-flow vertical flux of air is induced by a fan installed on the top of the tower to strip the volatile organic compounds (VOCs) out of the water. The air flow, reach in VOC concentrations, is then released to the atmosphere after proper treatment, while the treated water is dropped in a sump tank at the bottom of the tower.

The main technical characteristics of the evaporative tower are the following:

- Type: evaporative cooling tower PMS 10/260 (MITA™);

- Air flow rate: 45,000 m^3/h;

- Water maximum flow rate: 60 m^3/h;

- Useful volume: 6.5 m^3;

- Sump tank volume 1.4 m^3.

The extracted groundwater passes through a strainer mounted on the pumping-main and is subsequently sent to the suction line of the recycling pump where it mixes up with the treated water. The mixture of treated water and extracted groundwater is boosted to the top of the tower and sprayed through the nozzles onto the filling pack.

The high specific surface of the filling pack allows maximizing the contact between the water and the counter-flowing air. The effect of the air on the contaminants dissolved in the water is twofold:

1. The oxygen contained in the air flow oxidizes the metal ions dissolved in the water forming insoluble compounds which precipitate at the bottom of the sump tank;

2. The air flow, forced by the top fan, stimulates the evaporation of organic contaminants from dissolved phase into gaseous phase (process of stripping).

The treated water is recovered in the sump tank beneath the tower and is then partly re-circulated and partly skimmed over into a settling tank located nearby the tower at a lower level. From there, the water is directly discharged into the facility sewer system.

For the setting up of the matrix based approach, two field tests with different operating conditions have been carried out on the evaporative tower.

The first field test consisted of a static trial run of the system with the objective of calculating the values of the coefficient k_i for three main typologies of contaminants (iron, manganese and VOCs). During this first test the following activities were carried out:

- The sump tank of the tower was filled with groundwater extracted from the well that presented the highest concentrations of contaminants;

- The system was turned on and let run in closed circuit to measure water flow rate, the head loss of the nozzles and then calculate the duration of each cycle of treatment;

- Samples of water were collected before the starting of the system (t_0) and at the end of each cycle (t_n with $n > 0$);

- Samples of water were collected to be analysed for Fe, Mn and VOCs both on site (by means of a photometer and a photo ionisation detector- PID) and in laboratory for chemical analyses.

A scheme of the system layout and process adopted for the first field test is displayed in Fig. **3**.

Once the system was turned on, the manometer and the flow meter installed on the recirculation pipeline allowed the following readings:

- Flow rate of the recirculation pump (that is the flow rate of the water treated by the system): $Q_t = 16.2$ m^3/h;

- Head loss of the nozzles $\Delta p = 2.5$ bar.

Given the volume of the sump tank (1.4 m^3) and relating it to the recirculation flow rate, the contact time was calculated and resulted in approximately 5 minutes. During this contact time, the water collected in the sump tank completes one cycle of treatment.

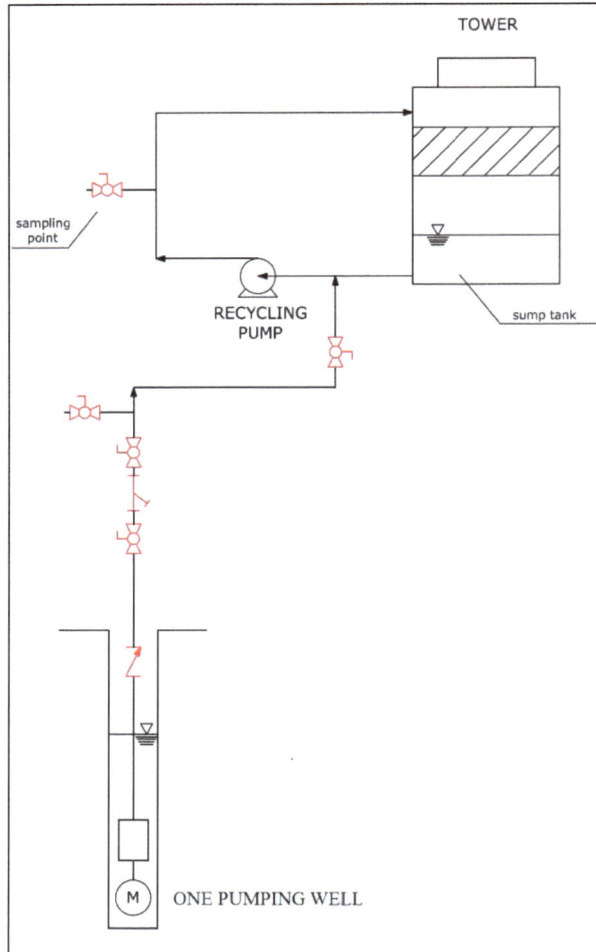

Figure 3: Layout for the field test 1.

The field test lasted almost two hours during which a total of 10 water samples (including t_0) were collected for *on site* and laboratory analyses.

Table **1** shows the results obtained from the laboratory analyses for Fe, Mn and TPHs.

Table 1: Chemical results – field test 1

Sample	Elapsed Time (min)	Fe (mg/l)	Mn (mg/l)	TPH (mg/l)
Analytical Method		EPA 6020A 2007	EPA 6020A 2007	EPA 5030 C 2003 + EPA 8260 C 2006
C00	0	2.9	3.5	49.66
C01	8	2.4	3.42	5
C02	16	1.1	1.18	0.69
C03	25	1.4	0.96	0.125
C04	35	0.06	0.78	0.005
C05	45	0.17	0.64	0.005
C06	55	1.1	0.57	0.005
C07	70	0.6	0.44	0.005
C08	85	0.19	0.355	0.005
C09	110	0.3	0.274	0.005

As displayed in Fig. **4**, after approximately 8 minutes (2 cycles of treatment) the concentrations of the considered contaminants went below the local regulation limits for discharging into surface water (respectively 2 mg/l for Fe and Mn and 5 mg/l for TPH): t=8 min can be considered as the residence time of the system.

Moreover, Fig. **4** shows that after about 35 minutes the concentrations of the considered contaminants reached an asymptotic trend.

The results of chemical analyses carried out on the water samples collected during the first field test confirm the assumption that a first-order kinetic can represent the contaminant concentration depletion through the treatment process.

As a matter of fact the data in Fig. **4** can be described by the following trend-lines:

- $c=2.9e^{-0.03t}$, for Fe concentrations;

- $c=3.5e^{-0.0286t}$, for Mn concentrations;

- $c=49.66e^{-0.2547t}$, for TPH concentrations (in this case, only values above the instrumental limit of 0.005 mg/l have been considered).

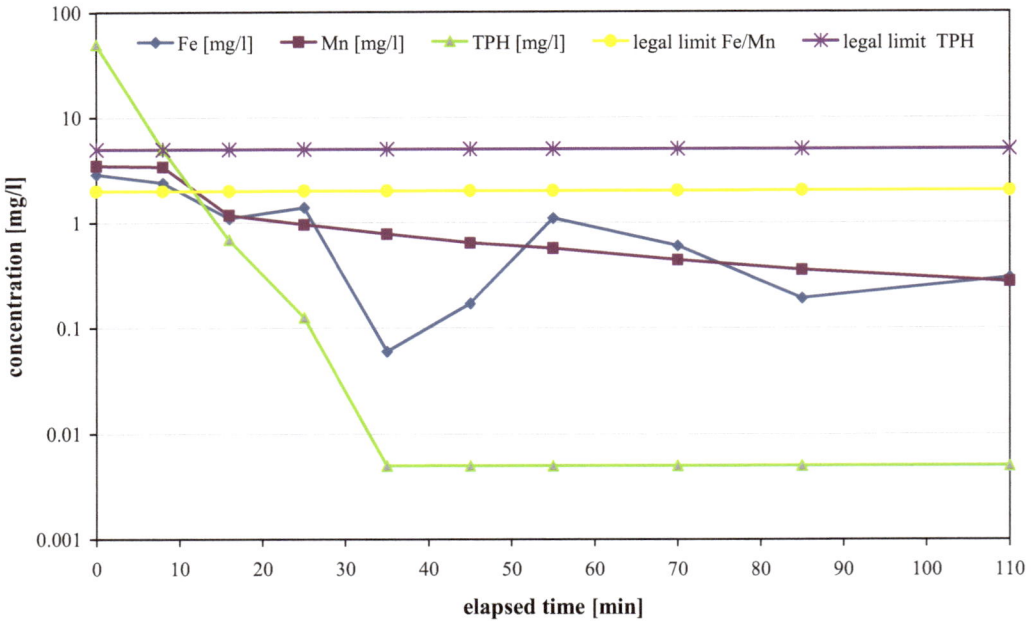

Figure 4: Contaminant concentrations *vs.* time within the cycled water.

Therefore it has been possible to consider the following coefficients k_i of each contaminant to be computed in the treatment matrix:

- $k_1 = 0.003$ min^{-1} for Fe;

- $k_2 = 0.0286$ min^{-1} for Mn;

- $k_3 = 0.2547$ min^{-1} for TPHs.

The second pilot test was carried out with the aim of verifying the functioning of the system under full operating conditions. During this test, whose functioning scheme is reported in Fig. **5**, the following activities were carried out:

- 3 pumping wells were turned on to fill the sump tank of the tower;

- Groundwater samples were collected from each pumping well before the starting of the system (t_0) and sent to the laboratory for chemical analyses;

- Once the sump tank of the tower was completely filled, the pumping wells were stopped and the system operated in closed circuit (only recirculation the water filling the sump, with no discharge) in order to build up a suitable amount of treated water to run the test;

- The 3 pumping wells were turned on again and tuned at the operating flow rates summing up to 5.5 m^3/h;

- Once the tower reached steady conditions (inlet flow rate equal to the outlet flow rate), the recirculation flow rate was tuned to a value determined applying the results of the first field test and equations (1), (2), (3) and (4), and with an overall inlet flow rate of 5 m^3/h, by operating the valve mounted on the discharge pipe as to keep constant the level of water in the sump;

- At given intervals of time (20, 30 and 40 minutes), 9 water samples (including 3 at t_0, one for each pumping well) were collected at the outlet to be analyzed.

The second pilot test lasted approximately two hours, with a measured recirculation flow rate (R) equal to 19 m^3/h, against the 16.2 m^3/h measured during the first test. This enhancement was likely due to a reduced head loss in the spray nozzles as a consequence of their use after years of inactivity. Despite the higher recirculation flow rate, the contaminants concentrations trend never reached an asymptotic value likewise the first order kinetic trend, but continuously fluctuating through time. As such these results have not been considered in this study.

The failure in knocking down contaminants concentrations was likely due to a mistake made in the system set up. In order to simplify the execution of the test, groundwater from the barrier wells was directly pumped into the sump tank in proximity of the suction main thus not allowing an efficient mixture with the treated water. As a consequence, the final setup design of the system was modified in order to avoid this malfunctioning. Nevertheless the second pilot test can be considered successful as it proved the functionality of the system also under a continuous feed set up, as it was tested in full operating conditions.

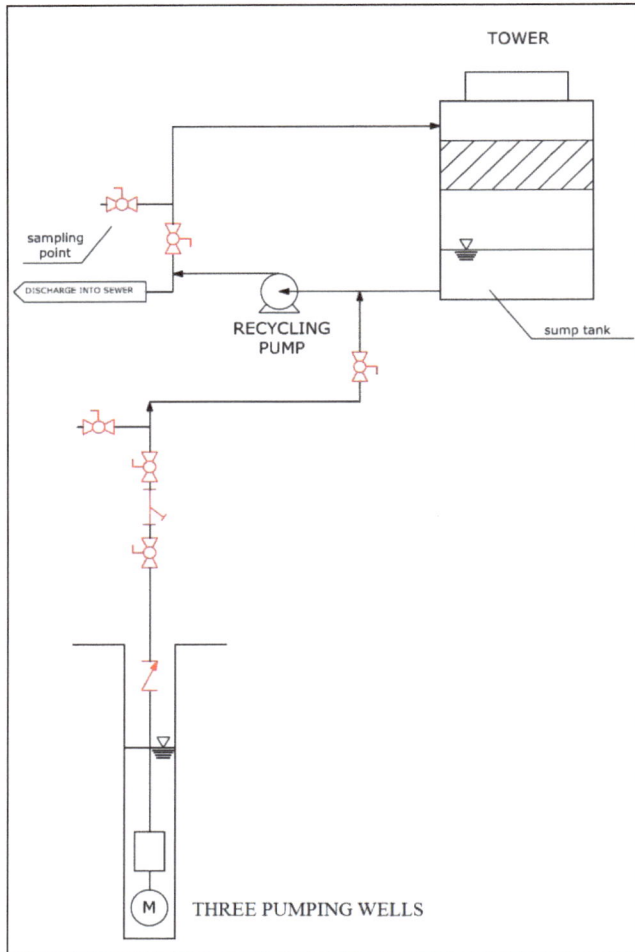

Figure 5: Layout for the field test 3.

RESULTS

In order to apply the proposed matrix based approach, it was necessary to carry out also pumping tests in order to determine the extraction flow rate for the hydraulic containment of the contaminated plume within the site. Under operating conditions of the P&T system, it has been determined a flow rate of 13 m^3/h of groundwater to be extracted from the pumping wells.

With the obtained set of data, it has been possible to apply the proposed matrix approach for determining the recirculation ratio that allows obtaining an output whose concentrations respect local regulation limits for discharging into surface water for the considered contaminants.

In particular, for the mathematical model, the elements of the treatment matrix have been obtained considering the values of the coefficients k_i and the residence time resulting from the findings of the first field test that was carried out. Consequently, for the three considered contaminants (Fe, Mn and TPHs), the 4×4 matrix M is equal to:

$$M = \begin{vmatrix} 0.79 & 0.00 & 0.00 & 0.00 \\ 0.00 & 0.80 & 0.00 & 0.00 \\ 0.00 & 0.00 & 0.13 & 0.00 \\ 0.21 & 0.20 & 0.87 & 1.00 \end{vmatrix} \tag{5}$$

Finally, by iterative matrix calculations of the four array relations (1), (2), (3), (4), a recirculation ratio R=66% has been proved to reach the local regulation limits for the three considered contaminants.

Therefore the treatment system was set up with the following operating parameters:

- Overall inlet flow rate from barrier wells (I): 13 m^3/h;

- Recirculation water flow rate (S): 25 m^3/h;

- Tower outlet flow rate (T): 38 m^3/h;

- Overall outlet flow rate (O): 13 m^3/h.

CONCLUSIONS

In the present research work, it has been investigated the possibility of adopting a matrix based approach for modelling a treatment process for contaminated groundwater. In particular, it has been considered a modified Broadbent and Callcott breakage-matrix whose elements have been replaced by first-order kinetic equations representing the depletion rate of contaminants' concentrations dissolved in the water. The adopted approach allows determining the recirculation ratio for groundwater in a treatment system in order to reach the water quality objectives for discharge.

In order to set up the mathematical model, two field tests have been carried out and afterwards the matrix approach has been applied for the revamping of an evaporative tower for the treatment of contaminated groundwater for three main typologies of contaminants (iron, manganese and volatile organic compounds).

The results of chemical analyses carried out on samples of groundwater, collected in the input and in the output flows of the evaporative tower, confirmed the effectiveness of the proposed approach for adjusting the treatment system performances modifying the recirculation ratio of treated groundwater before being discharged.

Further investigation could be carried out in order to refine the model and to test a predictive use of the matrix approach for estimating the concentrations of contaminants in the output of wastewater treatment systems.

ACKNOWLEDGEMENT

Declared none.

CONFLICT OF INTEREST

The author(s) confirm that this chapter content has no conflict of interest.

REFERENCES

Broadbent, S.R., Callcott T.G. (1956). *Philosophical Transactions of the Royal Society Lond.* A 1956 249, 99-123.
Cohen, R.M., Mercer, J.W., Greenwald R.M., Milovan, S. (1997). *Design Guidelines for Conventional Pump-and-Treat Systems*, Beljin, EPA/540/S-97/504.
Endres, K.L., Mayer, A., Hand, D.W. (2007). Equilibrium *versus* Non-equilibrium Treatment Modeling in the Optimal Design of Pump-and-Treat Groundwater Remediation Systems. *Journal of Environmental Engineering 133*, 809.
Javandel, I., Tsang, C.F. (1986). *Capture-zone type curves: a tool for aquifer cleanup.* Groundwater, *24*(5), 616-625.
U.S. EPA (1990). *Basics of Pump-and-Treat Ground-Water Remediation Technology.*
U.S. EPA (2008). *A systematic approach for evaluation of Capture Zones at Pump and Treat Systems*, EPA/600/R-08/003

Send Orders of Reprints at reprints@benthamscience.org
Separating Pro-Environment, 2012, 139-171

Separation Technologies for Inorganic Compounds Contained in Industrial Wastewaters Including Metal Ions, Metalloids, Thiosalts, Cyanide, Ammonia and Nitrate

N. Kuyucak[1,*] and I. Toreci Mubarek[2]

[1]*Golder Associates Paste Technology Ltd., 32 Steacie Drive, Kanata ON, K2K 2A9, Canada and* [2]*Golder Associates Ltd., 32 Steacie Drive, Kanata ON, K2K 2A9, Canada*

Abstract: Industrial wastewaters such as mining effluents and landfill leachate may contain metal ions in higher concentrations than those for which the regulatory standards require a treatment for their removal/reduction. Chemical precipitation/settling is the most common method used for removal/reduction of metal ions (*e.g.* iron, lead, copper, nickel, zinc, aluminum, manganese, *etc.*). Separation of metalloids (*e.g.* arsenic, selenium, molybdenum, antimony, *etc.*) can be achieved by co-adsorption onto iron or aluminum. Thiosalts, incomplete anoxysulphur compounds and cyanide often require the addition of a strong oxidation reagent. Biological nitrification and denitrification are conventional methods for removal of ammonia and nitrate. Respectively as industry accepts and often prefers to use conventional methods as proven and best available technology economically achievable (BATEA), depending on the site conditions and requirements. Emerging technologies such as membrane technologies (*e.g.* nanofiltration, reverse osmosis, *etc.*) have only been recently applied to many sites. Innovative approaches, such as snow making, are also used to separate ammonia and metal ions from wastewaters. Innovations have progressed on developing either new technologies or new approaches for mode of application of known technologies. Mode of application could vary from active methods, where pumps and pipes are used and the process takes place in reactors under controlled conditions, or passive methods, where the use of pipes and pumps is eliminated or limited. Wetlands (natural or engineered) and peat filters are often considered as passive methods. Conventional, emerging and innovative separation technologies and application of passive and active methods as well as capabilities and limitation of processes are discussed.

Keywords: Heavy metals, Metalloids, Ammonia-nitrate removal, Cyanide destruction, Thiosalts, Landfill leachate treatment, Mining effluents, Acid mine drainage, High density sludge process, BATEA, Chemical precipitation, Lime neutralization, Passive treatment, Biological treatment, Sulphate reducing bacteria, Wetland, Peat filter, Waste management, Ion exchange, Membrane technologies.

INTRODUCTION

Mining and metallurgical, waste management, food and beverage, pulp and paper,

*Address correspondence to N. Kuyucak: Golder Associates Paste Technology Ltd., 32 Steacie Drive, Kanata ON, K2K 2A9, Canada; Tel +16135929600; Fax: +16135929601; Email: Nural_Kuyucak@golder.com

textile, and pharmaceutical industries are examples of activities that very often produce wastewaters or effluents that may require appropriate treatments for separation of metal ions or compounds before they can be safely released to the environment. Mining activities, starting from exploration to decommissioning stages, generate effluents requiring proper handling. Effluents from mining and metallurgical processes and waste management activities (*e.g.* landfill leachate) show a more complex nature in comparison to effluents originating from other industries because they could contain acidity, metal ions, metalloids, thiosalts, cyanide and/or nitrogen compounds, ammonia and nitrate. These effluents have to be treated for removal/reduction of undesirable compounds. Even after their closure, mining and waste management (landfill) sites often require effluents to be collected and treated before it is possible to discharge them to the environment. The water quality objectives and regulatory standards are key factors for the selection of the technology that results as the most appropriate treatment option for a given site. The common approach is to assess the best available technologies (BAT) and/or the best available technology economically achievable (BATEA) for the considered site. Information on the sources and composition of mining and metallurgical effluents and landfill leachate and treatment objectives, including regulations and guidelines used in different countries, are described below.

SOURCES AND CHARACTERISTICS OF MINING EFFLUENTS

At a mining site, disturbances to the ground start with drilling activities in the exploration phase and continue until the site closure. Generation of effluents to be treated may start from the beginning of mining activities and may continue for many years after the decommissioning of the site. Potential sources of mining effluents that could require treatment are described in the following paragraphs.

ACID MINE (OR ROCK) DRAINAGE (AMD OR ARD)

The management of waste materials, such as tailings and waste rocks coming from mining of sulphidic metal ores, and sulphidic uranium and coal ores, represents an environmental challenge for mining companies. Acid generation occurs when sulphide minerals (predominantly pyrite - FeS_2 - and pyrrhotite - FeS) contained in the waste material are exposed to oxygen and water (Kuyucak, 2001a, b; Kuyucak, 2002). Leaching of acid products can happen as rainwater and/or snowmelt enter the waste pile or dump. If sufficient alkaline or buffering minerals (*e.g.* calcite) are not present to neutralise the acid, the resulting leaching water becomes acidic. The solubility of many metals is greatly increased under low pH conditions. Therefore the increased acidity in leaching water can result in high dissolved concentrations of metal ions such as iron (Fe), manganese (Mn), aluminum (Al), zinc (Zn), copper (Cu), nickel (Ni), lead (Pb), cadmium (Cd), arsenic (As), *etc.* This leaching water is generally known as acid rock drainage (ARD) or acid mine drainage (AMD). According to its acidity and metal content, AMD can be classified as low, medium, or high strength (Kuyucak *et al.*,1995). Acid generation can also be observed on roads, bridges and tunnels where sulphidic rock is used for construction.

MINE DEWATERING – AMMONIUM AND NITRATE

Explosives made of ammonia (NH_4-N) and nitrate (NO_3) compounds are in general used wherever blasting is required for extraction of the ores. As a result, water produced from mine dewatering may contain NH_4-N and NO_3 in elevated concentrations (Kuyucak, 1998). Careful management and proper disposal are necessary to prevent impacts to surface water and/or ground water resources.

PROCESS WATERS – CYANIDE AND ACIDS

Chemicals such as cyanide, sulphuric acid, and hydrochloric acid are used for processing ores and recovering metals. Cyanide (CN) is widely used for extracting precious metals (*i.e.* gold, silver, and platinum) as well as for conditioning mineral processes (*e.g.* flotation) to improve recovery of base metals such as Cu, Pb, and Zn (Kuyucak, 1998).

MINERAL PROCESSING AND TAILINGS RECLAIM WATER - THIOSALTS

Grinding (milling) and flotation of complex sulphide ores in an alkaline media produce a series of sulphur oxyanions called "thiosalts". Thiosalts include thiosulfates, polythionates, sulphide, and sulphate. They represent a delayed acid-generating capacity in mill effluents that results in a decrease in pH and, subsequently, an increase in metals and total suspended solid (TSS) concentrations. Treatment of mill effluents to remove/reduce thiosalts is necessary as current technology is not able to prevent the production of thiosalts in the grinding and flotation circuits (Kuyucak *et al.*, 2001; Kuyucak, 2001c).

TREATMENT OBJECTIVES AND APPROPRIATE PROCESS SELECTION

Mining effluents may require treatment by means of physical, chemical, and/or biological separation processes to minimize or eliminate their potential impacts on the environment. Undesirable compounds such as acidity, metals, ammonia/nitrate, cyanide, thiosalts, and TSS have to be removed or reduced to concentrations complying with regulatory standards for water quality. Treatment objectives are usually set based on the local regulatory standards, which also influence the selection and design of an appropriate treatment process for a given site.

Regulatory Standards

Each country sets standards to regulate the quality of water discharged in the environment and of the receiving water reservoirs. Regulatory standards for water quality in Canada are set at federal and provincial or territorial levels of government and involve the monitoring of certain parameters in waters. For instance, in the 1990s, the Ontario government launched the Municipal Industrial Strategy for Abatement (MISA) program for the management of "end-of-pipe" industrial discharges of potentially toxic substances into Ontario's waterways (the metal mining sector is just one of the nine industrial sectors covered by MISA). MISA sets daily and monthly maximum concentrations for certain parameters and requires that the end-of-pipe effluent must not be toxic to aquatic organisms. At this aim, specifically rainbow trout fry and water fleas (*i.e. Daphnia magna*) are tested. Mining sites located in the Province of Quebec must meet the water quality standards described in MENVIQ Directive 019. Alternatively, site specific standards (*e.g.* Policy 2: No more than background value) can be set by Quebec authorities for a specific site (Environment Canada 2001, Ontario Drinking Water Standards, 2000).

Additionally, in 1998, the Canadian provincial governments (except Quebec) prepared a Canada-wide Accord on environmental harmonization, aimed at achieving the highest level of environmental quality for all Canadians. Through the Canadian Council of Ministers of the Environment (CCME), provincial ministers set priorities and established work plans for addressing Canada-wide significant issues and for implementing the commitments set out in the Accord on a partnership basis. CCME's water quality task group has developed water quality guidelines mainly for the protection of aquatic life and for agricultural water uses. Site specific objectives can be set by the authorities depending on the site requirements.

The World Bank and the World Health Organization also set standards for waters resulting from mining activities; their standards are particularly applicable for the projects that take place in developing countries, or which depend on World Bank or IFC financing (World Bank, 1997).

Sustainability Considerations

Treatment processes can be designed to produce effluent and sludge of a quality that allow their recycling and reuse for other purposes, including mining processes, agricultural water for irrigation and livestock, recreational water, hydro-electric generation, and process water for industries located in the vicinity of the mining site.

At the Kingsmill Tunnel site in Peru in collaboration with the Peruvian Water Works Company (SEDAPAL), AMD has been treated using a high density sludge (HDS) process in order to use treated water as an additional source of drinking water for the City of Lima. Within pilot studies, physical, chemical, biological, and toxicological tests were conducted on the treated water and the results of analyses were compared with both the Peruvian Drinking Water Quality Standards and the U.S. EPA standards. The treated AMD was found to be acceptable as an additional source to primary drinking water resources (*e.g.* lake and river water). The process was also found to be economically viable as it was able to generate revenues from the selling of the treated AMD (Kuyucak *et al.*, 2003; Kuyucak *et al.*, 2004).

Boliden implemented a HDS process at the Apirsa site in Spain in order to obtain a water quality that could be recycled back to the mining/mill processes, thereby reducing the need for fresh water use (Kuyucak *et al.*, 1999).

Recovery of valuable metal hydroxide sludge and/or metal sulphide sludge has been investigated (Hedin, 2006; Kuyucak *et al.*, 1994; Rao *et al.*, 1994). Metal laden wastes can potentially be sent to smelters or metal manufacturing processes as a secondary feed material. In addition, the use of sludge as a back fill material has been put into practice (Kuyucak *et al.*, 2001b).

POTENTIAL QUALITY OF WATER FROM MINING, MILLING AND METALLURGICAL PROCESSES

Mine water quality profiles vary according to the nature of the exposed minerals and ore bodies. A large spectrum of quality profiles, that can be found in water resulting from mining/metallurgical processes could include: alkaline and saline waters; acid, saline and metals rich waters (Kuyucak, 2006). Salinity profiles typically consist of anions (*e.g.* carbonate, sulphate, chloride) and cations (*e.g.* sodium, calcium, magnesium, and metal ions such as Fe, Al, Mn, Zn, Ni, Cu, *etc.*).

Due to the use of explosives, effluents resulting from dewatering of open pits and/or of underground mines may contain ammonia and nitrate. CN may also be found in waters coming from metallurgical processes through its use as a chemical reagent in base metal mines and in leaching reagent in gold/silver (precious metals) mines. Degradation of CN may also result in ammonia and nitrate compounds. Incomplete oxidation of sulphides during milling and flotation processes may generate thiosalts (S_2O_3, S_2O_4, S_2O_6, *etc.*) which continue to oxidize until the final oxidation product (SO_4) is reached. This process leads to a delayed acidity for receiving water reservoirs, as oxidation of thiosalts generate H^+ ions (*i.e.* proton acidity).

SEPARATION TECHNOLOGIES

Best Available Technology Economically Achievable (BATEA)

Physical, chemical and/or biological methods can be adopted as treatment options depending on the type of contaminants found in the mining effluents. Common heavy metals such as Cu, Zn, Ni, Cd, Pb, and $Fe^{(2+/3+)}$ could be efficiently removed/reduced by lime neutralisation/precipitation process, which is accepted as a best available technology economically achievable (BATEA) and proven method. Acid is neutralised while metals (Me) and sulphate are precipitated respectively in the form of metal hydroxides and gypsum ($CaSO_4$), as shown in Equation 1. The mixture of precipitates is called sludge and requires proper management for its disposal and storage. Therefore, the amount and quality of sludge generated is an important parameter for the choice of the treatment option.

$$Ca(OH)_2 + Me^{2+}/Me^{3+} + H_2SO_4 \leftrightarrow Me(OH)_2/Me(OH)_3 + CaSO_4 + H_2O \quad \textbf{(1)}$$

Air is frequently used to oxidise ferrous to ferric iron during precipitation to obtain sludge that is more chemically stable. Due to the amphoteric nature of metal hydroxides, if the solubility limit for each metal is achieved at an optimum point, above and below the optimum pH, metal hydroxides become more soluble as shown by solubility curves in Fig. **1**. For instance, at pH 9-10, Cu and Zn can be reduced to levels of <0.1 mg/l, while Pb and Fe^{3+} are reduced to µg/l. Levels of Cd and Ni below 1 mg/l can be achieved with pH adjustment above 10 unless the water contains high levels of iron. Cr can be lowered to levels of <0.5 µg/l at pH 7-8, following reduction of Cr^{6+} to Cr^{3+}. Mn removal requires strong oxidation followed by liming at a pH greater than 10. Removal of metal ions by neutralization/precipitation is summarized in Table **1**.

When water contains metal and sulphate ions in low concentrations (*e.g.* Fe^{2+}<200 mg/l and SO_4<2,000 mg/l), caustic reagent and air could be injected into a port of a jet pump (in-line treatment). In this way, a single step process could be set up and subsequently savings in capital, operating and maintenance costs could be achieved.

Table 1: Removal of metal ions by neutralization/precipitation

Metal	pH	Concentration Achievable	Comment
Cr	7-8	<0.5 µg/L	reduction of Cr^{6+} to Cr^{3+}
Cu & Zn	9-10	<0.1 mg/L	
Pb & Fe^{3+}	9-10	µg/L range	
Cd & Ni	>10	<1mg/L	hindered by high Fe concentrations
Mn	>10	<1 mg/L	strong oxidation required

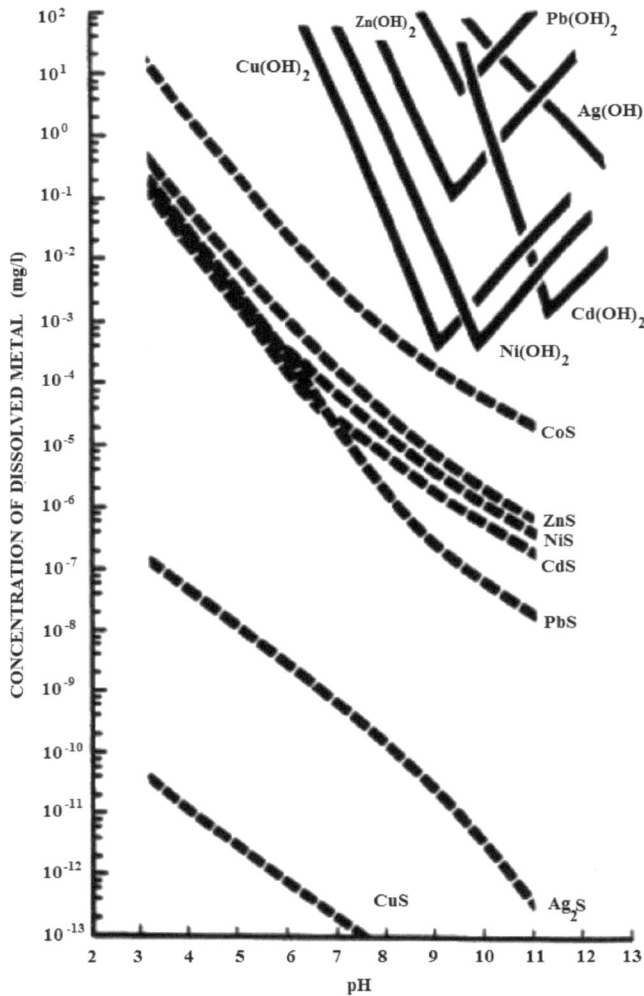

Figure 1: Relationship between pH and solubility of metal hydroxide and metal sulphide compounds (Kuyucak, 2007).

MODE OF NEUTRALIZATION/PRECIPITATION PROCESS APPLICATION – ACTIVE PROCESSES

Lime neutralisation facilities may range from the simple addition of lime to the tailings pipelines up to facilities consisting of reactors, clarifiers, and sludge dewatering equipment, as the High Density Sludge (HDS) process illustrated in Fig. **2**. Type III and IV processes, where a portion of the sludge generated is recycled from the clarifier underflow to the process and either it is used alone to partially neutralise the mining effluents or along with lime as the alkaline reagent, are capable of producing more compacted sludges (*e.g.* 10-30% solids) than traditional methods of liming. Flocculant can be added either in a (flocculant) reactor or in the pipeline or directly in the clarifier feed-well. HDS process usually removes SO_4 in high concentrations up to the theoretical solubility limit of gypsum. Gypsum is made of crystalline particles with a defined shape which allows dewatering of sludge to low moisture contents (*e.g.* 60% solids) (MEND, 1994; Kuyucak and Sheremata, 1995). Although HDS capital and operating costs are higher than the ones of conventional methods, the recent tendency in mining is to use HDS or to upgrade to HDS existing plants. Due to the improved sludge characteristics (*e.g.* less volume) and effluent quality (*e.g.* less SO_4), HDS offers a number of cost advantages (Zinck and Griffith, 2000). It increases the quantity of the recovered water and, subsequently, the quantity of lime used per unit of treated water decreases. Further, it allows increasing the water that can be recycled back to the process, thus reducing the need for fresh water at operations (Kuyucak *et al.*, 1999). Scaling in the process is significantly reduced due to removal of SO_4 in high concentrations. The process control is reasonably well automated requiring less maintenance and labour. Sludge disposal and site monitoring costs are lowered. Formation of gypsum crystals improves the paste properties when acid tailings are neutralized or sludge is mixed with tailings for paste production.

Separation of Metalloids from Effluents by Chemical Precipitation

Contaminants such as arsenic (As), antimony (Sb), molybdenum (Mo) and selenium (Se) are called metalloids because they are found in the form of anionic complexes in waters. They require the use of additional chemicals such as H_2O_2, $FeCl_3$ or $Fe_2(SO_4)_3$, or Al-based (alum) products or Na_2S and/or CO_2 as adjuncts to the lime process. Before the addition of co-precipitating reagents (*e.g.* Fe^{3+} product), the adoption of an oxidation process to oxidize As^{3+} to As^{5+} is often required to achieve a high level of separation. Hydrogen peroxide is a common oxidizing reagent and is used in conjunction with Fe products. The pH is an important parameter that plays an important role in the removal of As and of other metalloid ions from waters. The pH should be set to an optimal level to obtain the desired rate of removal. Co-precipitation with iron at a pH 3-4 could decrease the concentration of Mo down to <0.5 mg/l (Kuyucak, 1998).

Sulphide Precipitation

Sulphide precipitation (using sodium sulphide - Na_2S) has been used to treat wastewaters coming from metal finishing industries; it is not routinely used to treat AMD. Solubilities of metal sulphides are usually several orders of magnitude less than metal hydroxides (Kuyucak *et al.*, 1991a, b; Kuyucak and Payant, 1995). The solubilities of metal sulphide precipitates are shown in Fig. **1**. Na_2S, FeS, $(NH_4)_2S$, BaS or H_2S can be used as reagents. The use of sulphide precipitation results in better metal removal from those effluents which contain phosphate, ammonia, organics, surfactants, chelators and Cr^{6+}. Metal sulphide complexes offer some advantages over hydroxide precipitates because they are less voluminous. They are also more chemically stable because they are less susceptible to changes in pH as long as they are stored under anaerobic conditions. In addition to the noxious H_2S gas production from the system, the settling and separation of fine and colloidal metal precipitates from the treated water pose potential problems; solid/liquid (S/L) separation may require a filtering system such as sand filter. Due to its higher costs in comparison to the ones of lime neutralisation, the application of sulphide is limited to site specific conditions. Biologically generated sulphide precipitation processes have been investigated as an alternative treatment method (Yanful *et al.*, 1991; Kuyucak, 2000).

To meet stringent final effluent regulatory limits, in order to control dissolved metals and suspended solids, Na_2S and more sophisticated solid/liquid separation equipment such as filters (sand or fabric or microfiltration (MF) or nanofiltration (NF)) could be employed (Kuyucak and Payant, 1995). For instance, addition of few mg/l of Na_2S at neutral pH levels could lower Cd concentrations (*e.g.* <0.01 mg/l) without affecting the solubility of other metals (Fig. **1**). When pH adjustment alone using an alkaline reagent is ineffective, the use of combination of alkaline and sulphide precipitation methods can be considered to meet the required standards for water quality. Oxidation of As^{3+} to As^{5+} is necessary to remove As from effluent prior to lime, sulphide and/or ferric iron precipitation. A common method for removing Hg consists in sulphide precipitation and results in an effluent of 10-20 µg/l as shown in Equation 2.

$$Hg + S \rightarrow HgS \tag{2}$$

Lime Sludge or Tailings Pond

Mine Water (MW)   **Overflow**

pH ~ 10   **Sludge retained in the pond**
                                    %S ~ 1-2

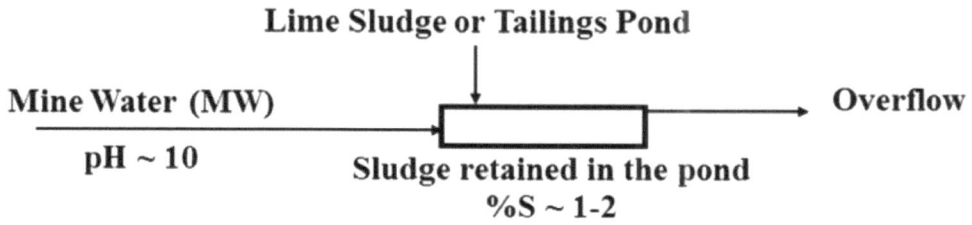

Type I – Lime addition to acid water, conventional method

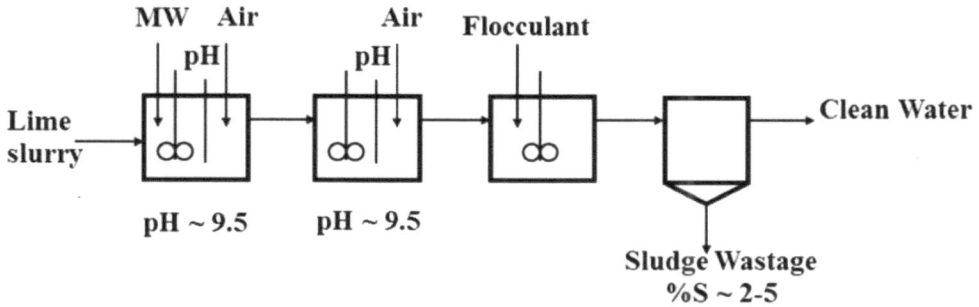

MW   Air   Air   Flocculant

pH| pH|

**Lime
slurry**   **Clean Water**

pH ~ 9.5   pH ~ 9.5

                                                    Sludge Wastage
                                                    %S ~ 2-5

Type II – Lime addition and neutralization reactors, conventional method

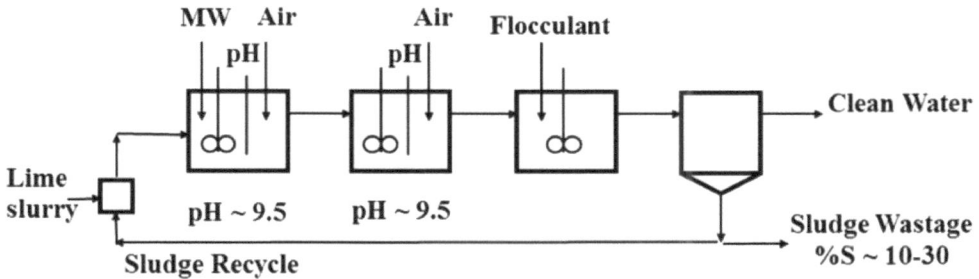

MW   Air   Air   Flocculant

pH| pH|

**Lime
slurry**   **Clean Water**

              pH ~ 9.5   pH ~ 9.5

Sludge Recycle   **Sludge Wastage**
                                                            %S ~ 10-30

Type III – Sludge recycle into lime slurry; High Density Sludge (HDS) process

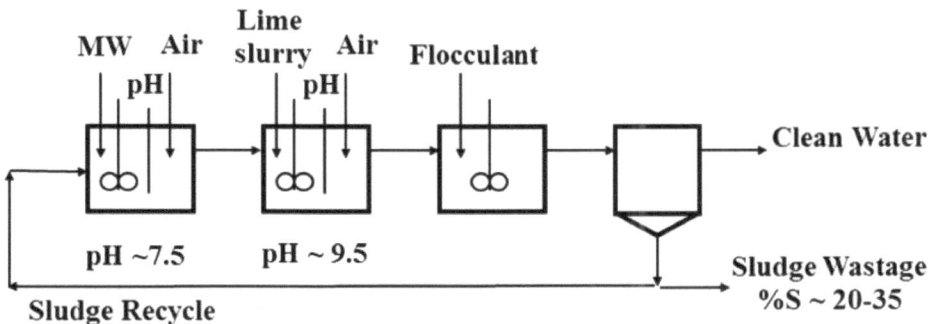

                        Lime
MW   Air   slurry   Air   Flocculant

pH| pH|

                                                                        Clean Water

pH ~7.5   pH ~ 9.5

Sludge Recycle   **Sludge Wastage**
                                                            %S ~ 20-35

Type IV – Sludge recycle into acid water; two-step HDS process

Figure 2: Types of lime neutralisation processes and resulting sludge densities.

Calcium Carbonate (CaCO₃) – Limestone Precipitation

Limestone ($CaCO_3$) can remove acidity and precipitate iron. Under controlled conditions, higher density sludges can be obtained by $CaCO_3$, as opposed to lime. By adding limestone in AMD, the $CaCO_3$ dissociates and CO_2 gas evolves, as shown below:

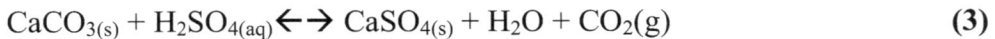

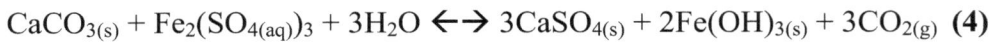

$$CaCO_{3(s)} + H_2SO_{4(aq)} \longleftrightarrow CaSO_{4(s)} + H_2O + CO_2(g) \tag{3}$$

$$CaCO_{3(s)} + Fe_2(SO_{4(aq)})_3 + 3H_2O \longleftrightarrow 3CaSO_{4(s)} + 2Fe(OH)_{3(s)} + 3CO_{2(g)} \tag{4}$$

The released CO_2 gas forms a carbonate ion which acts as a buffer and sets an upper limit for pH (max pH = 6.5) and also affects the rate and amount of lime consumption.

The precipitates can settle very slowly because of their small particle size. Removal of a broad range of metals and ferrous iron cannot be achieved since removal of many metals requires pH levels higher than 6.5. A treatment process involving the sequential addition of limestone and lime has been suggested for removal of a wide range of metal ions (Kuyucak, 1995).

Magnesium Hydroxide (Mg(OH)₂) Precipitation

When properly applied in the neutralization system, the addition of $Mg(OH)_2$ can result in a lower volume of more dense metal hydroxide sludge. $MgSO_4$, which is more soluble than $CaSO_4$, is formed in the process and $Mg(OH)_2$ can remove metals through surface adsorption. However, the rate of neutralization is slow and the buffering capability of $Mg(OH)_2$ prevents the pH from exceeding 9. Depending on the pH requirements, it can be used in conjunction with NaOH. $Mg(OH)_2$ is usually employed in treatment plants where the disposal costs of the generated sludges are high, such as in the Canadian Copper Refinery (CCR) in Montreal east, in order to reduce sludge disposal costs (Kuyucak *et al.*, 1991).

Sodium Hydroxide (NaOH) Precipitation

NaOH is highly reactive and results in less voluminous sludge. However, NaOH is expensive and the resulting sludge does not settle well, requiring filtering in most cases.

Ammonia (NH_3) Precipitation

Ammonia use is mainly preferred by coal mining industries due to its high solubility and to its properties that allow producing less sludge. Ammonia is usually injected as gaseous anhydrous ammonia near the bottom of ponds or inlet water (Kuyucak, 2000). Some hazards are associated with the handling of ammonia, and there are some uncertainties concerning potential biological reactions.

Waste or By-Products of Industries Precipitation Reagent

Some industrial wastes or by-products such as fly ash, kiln dust, paper pulp, and red mud (bauxite residue), deriving respectively from power plants, crude oil combustion gasification processes, paper production, and aluminum extraction, have the potential to be a lime substitute for the treatment of AMD. However, metal contaminants present in these compounds may raise some concerns. In addition, their neutralisation potential and reaction rates may be low and slow as compared to lime (Kuyucak, 2000; Zinck and Griffith, 2000).

Separation of Fine Particles and Total Suspended Solid (TSS)

Colloidal particles that cannot settle due to repulsive charges, and fine particles (fine precipitates) may result in high TSS that need to be aggregated to improve solid/liquid (S/L) separation or sedimentation process. Coagulation process leads to the formation of aggregates called flocs, which are compact and heavier particles that can settle faster. The addition of a coagulant, such as inorganic Al^{3+} or Fe^{3+} salts or organic 'polymers', helps to discharge or destabilise the electronegative colloids and serve as a bridge for neutral particles. Addition of a flocculation reagent (*e.g.* polymer) to coagulated particles further improves the aggregation process by bringing the particles together. Coagulation and flocculation reagents are often used together in waters. The critical parameters affecting the performance of the process include: the type of polymer, temperature of the system, viscosity and chemical characteristics of the pulp, and external stirring (Kuyucak, 2006). As fish gills are negatively charged, the use of cationic polymers (or their residual quantities in water) may result in some concerns.

A basin or a clarifier (*e.g.* regular circular clarifier or a lamella) can be used for the settling/separation processes. The use of a filter and/or a membrane process is also considered in conjunction with a coagulation and/or flocculation process to obtain a better TSS separation.

Separation Methods for Sulphate (SO_4) From Waters

Several methods for removal/reduction of SO_4 from waters exist and they may include: precipitation with barium (Ba); ion exchange separation; Biological Sulphate Reduction (SRB process) and aluminum hydroxide ($Al(OH)_3$) precipitation. In general, these methods are not widely used for the treatment of mining effluents. Even if they show technical feasibility, their application may not be economically feasible especially for large quantity of flows. The SRB and $Al(OH)_3$ precipitation processes are still under development. The methods, their capabilities and limitations are briefly discussed below.

Precipitation with Barium (Ba)

SO_4 ions can be precipitated with Ba as $BaSO_4$ using $BaCl_2$. The removal of precipitates from the resulting effluent requires sophisticated solid/liquid separation methods (*e.g.* sand filters). Otherwise, residual Ba-precipitates in the treated water could pose toxicological concerns since Ba is known to be toxic to aquatic life. In addition, SO_4 ions will be replaced with Cl ions which are also regulated and more potentially toxic to the aquatic environment than that of SO_4 ions.

Ion Exchange

Synthetic ion exchange resins are made of a polymer containing exchangeable functional groups. They are used for demineralising waters by removing cations and anions, such as in water softening. They are classified on the basis of their functional groups including strong-acid cation exchangers, weak-acid cation exchangers, strong-base anion exchangers and weak-base anion exchangers. Weak-base exchangers can remove free mineral acids such as HCl and H_2SO_4. Strong-base resins have been operated for removal of nitrate and sulphate from municipal water supplies at wide pH ranges, which could be suitable for removing SO_4 from the treated AMD. When the resins become exhausted (*i.e.* loaded with SO_4), they have to be regenerated by washing with NaOH or NaCl for their reuse. Regeneration of NaOH may cause an increase in pH in the final effluent requiring the addition of CO_2 into the treated water for final pH adjustment. Regenerating the resins with NaCl is practiced with caution, because the concentration of Cl in the treated (final) effluent may result in higher concentrations than the ones indicated by regulatory standards. In addition to high capital and operational costs of the process, the blinding of the resin by Ca and Mg ions, present in the water, and the disposal of the resulting brine (*i.e.* concentrated solution of SO_4) pose also some concerns.

Biological Sulphate Reduction (SRB)

A group of bacteria, called SRB, in the presence of an organic nutrient can convert SO_4 to sulphide (S^{2-}) and generate bicarbonate alkalinity under anaerobic and reducing conditions. The generated S^{2-} removes metals (*e.g.* Cu, Pb, Zn, Fe) forming insoluble complexes with them, while the produced bicarbonate causes an increase in the pH. SRB have been used in both passive and active process applications. THIOPAQ is one of the patented processes developed based on SRB (INAP, 2003). Its functioning was demonstrated at the Kennecott mining site for removal of sulphate and recover of Cu from waste rock leachate. To date, the process has only been implemented at the Budelco B.V. (smelter) in the Netherlands for reducing SO_4 levels from 1,500 mg/l to less than 500 mg/l in groundwater. In this process, several reactors are operated under controlled conditions and ethanol is used as nutrient for the bacteria (Kuyucak, 2000). Metals are removed achieving low concentrations, while excess sulphur (S), that remains after removal of metals, is converted to elemental S. In active applications, usually, ethanol or natural gas is used as nutrient source for the bacteria. As the nutrient source in the passive applications can be wastes or by-products coming from other industries, SRB applications in passive mode have received more attention. However, passive applications cannot handle the treatment of large flows.

$Al(OH)_3$ Precipitation

A South African research organization, Mintek, has been developing a process to convert AMD into potable water by removing/reducing SO_4 concentrations (Smith, 2000). The process is called SAVMIN and consists of several precipitation reactions (INAP, 2003). The first stage raises the pH using lime, and this precipitates heavy metals and magnesium as hydroxides. The hydroxides, including supersaturated calcium sulphate (gypsum), are then separated with the help of a clarifier/thickener.

In the third stage, aluminum hydroxide is added to the solution, and this causes the formation of the insoluble salt ettringite, which removes calcium and sulphate. The solution obtained at this stage is then treated with carbon dioxide to lower the pH, and the resulting pure calcium carbonate is removed from the water by filtration. The final stage of the process is the recycling of ettringite which is treated with sulphuric acid to regenerate aluminum hydroxide. For an application where the SO_4 concentration and flow rate were respectively about 2,200 mg/l and 4,167 m^3/h, capital and operational costs are respectively estimated to be about R 297 million (or US\$ 35 million) and R 0.07/m^3.

Emerging Technologies

Emerging technologies aims at achieving better separation of contaminants from waters in order to meet more stringent quality standards for treated water. They are used either as an alternative to conventional methods to replace them (*e.g.* chemical precipitation processes) or in conjunction with the conventional method depending on the site requirements. They are mainly based on physico-chemical processes (*e.g.* ion exchange, solvent extraction, membrane processes – that is, reverse osmosis (RO), ultrafiltration, MF and NF, and silicate micro-encapsulation). Their use is limited to site-specific applications because they are still not technically and economically feasible. For instance, several mining sites have been recently employing RO process for removal of total dissolved solids (TDS) from waters (Allen, 2008). Biological processes such as wetlands, the above described SRB, nitrification/denitrification or chemical co-precipitants (*e.g.* permeable reactive barriers "PRB") are relatively new technologies that find applications at mining sites. In a PRB system, the reactive material usually consists of zero-valent-iron (ZVI, elemental iron), limestone and organic materials. ZVI is an electron donor used to reduce metal contaminants such as Cr^{6+} and As^{5+}, dehalogenate hydrocarbons and precipitate (oxy)anions from ground water (INAP, 2003). Passive processes such as wetlands, SRB processes, anoxic limestone drains (ALD) and biosorbents are also considered as emerging methods (Kuyucak and Palkovits, 2009). In general, passive processes have been found to be suitable for the treatment of low flow and low-strength (that is, metal concentrations) wastewater situations. However, ALD have been used widely in the US to restore abandoned mines and for pretreating acid drainage before routing to a constructed wetland.

Ion Exchange Resins

Ion exchange (IX) appears to be an alternative method to remove/reduce metal ions (*e.g.* Cu, Ni, Zn, Hg, Mo) from waters containing single ions and/or simple ion matrix. IX can also be used as a polishing step. IX can achieve low levels of Hg and Mo, respectively 1-5 µg/l and 2 mg/l. Resin blinding due to presence of multiple ions in the water may cause inefficient use of IX. Separation of contaminants by IX resins is usually facilitated in columns that could either be operated up-flow or down-flow mode. Once the IX resin becomes exhausted (or saturated, *i.e.* reach "breakthrough" point), with the help of a chemical reagent, the resin is eluted and recovered for its reuse. The contaminants (*e.g.* metal ions) that are obtained in a concentrated solution are recovered or disposed of.

A continuous IX process (Sulf-IX™) was tested at a pilot-scale for treating industrial wastewater with TDS levels in excess of 20 g/l, mainly composed of ammonium sulphate and calcium sulphate (Bioteq, 2010). The cation exchange circuit selectively removed calcium from the water using strong acid cationic ion exchange resin. Then, the resin was regenerated using sulphuric acid. The anion exchange circuit selectively removed sulphate from the plant feed by mean of a weak base anion resin, which was finally regenerated using lime.

ZVI is also used as IX resins and adsorbent materials in columns. ZVI recently have received attention for removal of As from mining waters. However, the exhausted iron cannot be recovered and is usually discarded. During the co-precipitation of contaminants, Fe dissolves and the treated water becomes rich with Fe which then may require further treatment.

Membrane Processes

The use of membrane processes including MF, NF and RO has been increasing as the technology has been advancing in reducing the energy requirements and in improving the quality of membrane materials. RO can produce fresh water with low TDS and, as waste, a brine solution that requires proper management. An RO system consists of: a pretreatment process; a high pressure pump; membrane assembly; and post treatment (INAP, 2003). Pretreatment process is necessary to protect membrane from scaling and fouling. A pretreatment process train generally involves chemical precipitation, settling separation and filtration, either with conventional methods or with membrane including MF and NF. Post treatment is applied to condition the treated water to obtain the required final water quality standards, such as pH, alkalinity, hardness, removal of H_2S and preparation for distribution.

RO has been recently often used for removal of TDS when recycle/reuse of water is required and fresh water resources are limited, especially in Africa, Australia and Europe (Gunther *et al.*, 2006; van Niekerk, *et al.*, 2006). Ion exchange and reverse osmosis generate a concentrated solution of contaminants (*i.e.* brine). Management of resulting brine poses concerns and requires proper disposal (or if possible reuse). Before disposal, evaporation is one of the methods used for separation of water from brine. In some cases, brine can be discharge into underground.

Removal/Reduction of Ammonia and Nitrate

The removal of nitrates/nitrites is usually achieved by biological denitrification methods that allow their reduction to nitrogen gas. Free ammonia can form soluble amine complexes with heavy metals such as copper, zinc, silver and nickel. Therefore, the presence of ammonia could inhibit the precipitation of these metals at values of pH above 9, which is known to be an effective range for the precipitation of metal hydroxides (Kuyucak, 1998; Kuyucak, 2002). Nitrate is the end-product of the cyanide oxidation process and is produced as a result of the chemical or biological oxidation of ammonia. Nitrification and denitrification reactions are presented below in Equations (5) and (6).

Nitrification: $NH_3 + O_2 \rightarrow NO_3$ (5)

Anaerobic (for NO_3) Denitrification: $NO_3 + C_{org} \rightarrow N_2\uparrow + CO_2\uparrow$ (6)

During the nitrification process, ammonia is aerobically converted to nitrate (and nitrite) by *Nitrosomonas* and *Nitrobacter*. In the conversion process, approximately 4.3 mg oxygen (O_2) per mg of ammonia-nitrogen oxidized to nitrate are needed and a large amount of alkalinity is consumed: 8.64 mg HCO_3^- per mg of ammonia-nitrogen oxidized. The optimal pH range for this process is quite narrow being 7.5 - 8.6 (Metcalf & Eddy, 1991). Dissolved oxygen (DO) concentrations above 1 mg/l are essential for nitrification to occur. Otherwise, oxygen becomes the limiting nutrient and nitrification slows or ceases. Conversion of nitrate to nitrogen gas can be accomplished by several genera of bacteria under anaerobic (without oxygen) conditions in the presence of an organic carbon nutrient source. Alkalinity is produced during the conversion process and the optimal pH range lies between 7 and 8. The organisms are also sensitive to changes in temperature and the optimal temperature range is between 15 and 45 °C. Microorganisms exhibit P uptake above normal levels when exposed to an anaerobic zone followed by an aerobic zone (*i.e.* alternating aerobic and anaerobic conditions). Therefore, microbiological removal of P takes place under sequencing cycles of aerobic and anaerobic conditions such as the Bardenpho Process (Metcalf & Eddy, 1991).

Removal/Reduction of Cyanide

Due to its high affinity for gold and silver, cyanide is able to relatively selectively leach gold and silver from ores. Regulatory standards are applied to cyanide and cyanide compounds in wastewater streams. Therefore, residues and wastewater streams containing cyanide and cyanide compounds have to be treated to reduce the concentrations of total cyanide and free cyanide below the regulatory limits.

The nature of cyanide treatment processes may range from natural degradation in tailings impoundment (*i.e.* natural attenuation in surface ponds) to highly sophisticated plant applications. Natural degradation in tailings ponds has been the most commonly used treatment method in most mills (*e.g.* in Canada) for many years. Although natural degradation is still used for cyanide removal, in the last two decades, several processes including chemical, biological and electrochemical methods have been developed to either supplement or supplant the natural degradation (Nelson, *et al.*, 1998; Ripley *et al.*, 1996; Smith and Mudder, 1991; Smith and Mudder, 1994; Ritcey, 1989). The recent tendency is to use highly sophisticated and automated treatment plants for removal/reduction of cyanides and cyanide related compounds from the mining/mill effluents due to the increasing concerns of regulators and communities. These plants are generally installed with on-line monitoring device.

Chemical processes may consist of alkaline chloride oxidation (alkaline chlorination), hydrogen peroxide oxidation, Inco (SO_2/air) oxidation, Hemlo/Golden Giant (copper and iron sulphate) precipitation and acidification-volatilization-regeneration (AVR). Methods available for removing or destroying cyanide and associated process mechanisms are summarized in Table **2**.

Table 2: Methods for removing/destroying cyanide and process mechanisms

Method	Processes and Mechanisms
Natural Degradation (Collecting/Holding In Ponds)	• **Volatilization** • **Biodegradation** • **Oxidation**

Chemical Addition Under Controlled Conditions	• Oxidation Processes
	    SO_2- Air Oxidation
	    Alkaline Chlorination
	        - Chlorine gas
	        - Hypochlorites
	        - Electrolytic (*in situ*) generation
	    Ozonation
	    Hydrogen Peroxide Oxidation
	    Iron Sulphide Processes
	    Prussian Blue Precipitation
	    Bacterial Oxidation
	• **Conversion to Less Toxic Forms**
	    Thiocyanide conversion
	    Ferrocyanide conversion
	• Acidification/ Volatilization/Neutralization

Adsorption Processes	• Ion Exchange • Activated Carbon • Ion Flotation
Electrolytic Processes	• Cyanide regeneration • Cyanide destruction

Cyanides are biologically or chemically oxidised and are then converted to ammonia and carbon dioxide. The natural degradation of ammonia involves the transpiration of dissolved ammonia gas from the wastewater. Increasing pond area and pH enhance the removal. A biological process unique to Homestake Mining in South Dakota decomposes metal-cyanide complexes and efficiently oxidises cyanides to ammonia, which is further oxidised by bacteria ('nitrification') to nitrate. Cyanide oxidation and degradation reactions are illustrated in Equation

(7). Equations (5) and (6) given above for removal/destruction of NH_3 and NO_3 proceed after destruction of CN to NH_3.

$$CN^- + O_2 \text{ (biologically or chemically)} \rightarrow NH_3 + CO_2 \qquad (7)$$

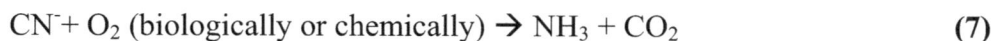

Base metal cyanide complexes are selectively oxidised to cyanate by a mixture of SO_2 and air (the 'Inco Method') in the presence of copper as a catalyst, in a controlled pH range. A number of operations use hydrogen peroxide to oxidize cyanides to cyanates. A premixed $CuSO_4/FeSO_4$ reagent can also remove cyanide at a pH of 9.5. CANMET-MMSL has tested cyanide recycling technologies as an economic and ecologic alternative to the destruction methods (NRCan, 2010). A pilot study was conducted for the evaluation of the AVR (Acidification Volatilization Reneutralization) process for a high copper effluent and the development of a non-aeration cyanide recovery method for a concentrated waste cyanide solution. Other methods such as air stripping, steam stripping, alkaline chlorination with hypochlorite at pH 10-11, engineered wetlands, acidification/volatilisation, adsorbents and ion exchange resins have limited use (Randol 1998; Kuyucak and Palkovits, 2009).

Removal/Reduction of Thiosalts

Grinding and flotation of pyrite (FeS_2) and pyrrhotite (FeS) ores in alkaline conditions produce a series of partially oxidized sulphur oxyanions such as thiosulphate ($S_2O_3^{2-}$), trithionate ($S_3O_6^{2-}$) and tetrathionate ($S_4O_6^{2-}$), which are collectively called "Thiosalts". Oxidation of thiosalts continues in waters until the end product of sulphate is reached. In this process protons (H^+) are also produced, representing a potential drop of pH in effluents or downstreams and a subsequent increase in metal and dissolved solids concentrations (Kuyucak, 2008). Examples of reactions for the oxidation and destruction of thiosalts are given in the Equations (8), (9) and (10), in Equation (11) by iron oxyhydroxides (FeOOH), and in Equation (12) by simply disproportionate:

$$S_2O_3^{2-} + 1/2O_2 + H_2O \rightarrow 2SO_4^{2-} + 2H^+ \qquad (8)$$

$$S_2O_6^{2-} + 2O_2 + 2H_2O \rightarrow 3SO_4^{2-} + 4H^+ \qquad (9)$$

$$S_4O_6^{2-} + 7/2O_2 + a\ 3H_2O \rightarrow 4SO_4^{2-} + 6H^+ \qquad (10)$$

$$S_2O_3^{2-} + 8FeOOH + 8H^+ \rightarrow 2SO_4^{2-} + 8Fe2^+ + 11H_2O \tag{11}$$

$$S_2O_6^{2-} + H_2O \rightarrow SO_4^{2-} + HS^- + H^+ \tag{12}$$

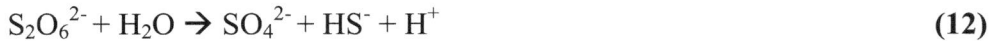

Thiosalts generation is site-specific. Currently, no method is available to cost-effectively prevent their production.

Thiosalts Separation Alternatives

Available alternative methods for thiosalts separation have been evaluated in laboratory or pilot scale tests and include (Vigneault *et al.*, 2003; Kuyucak *et al.*, 2001; Kuyucak and Yaschyshyn, 2007):

- Natural Oxidation in tailings ponds;

- Chemical Oxidation by hydrogen peroxide (H_2O_2), chloride (Cl_2) or ozone (O_3);

- Increasing Buffering Capacity by adding carbonate (CO_3^{2-}) and/or bicarbonate (HCO_3^-);

- Membrane and Electrochemical processes (*e.g.* RO), electrodialysis (ED) and electro-oxidation (EO);

- Air Oxidation (*e.g.* alkaline oxidation, Cu-catalyzed air oxidation and SO_2-air oxidation);

- Biological Oxidation using rotary biological contractors, rock filters and activated sludge;

- Biological Reduction to sulphide followed by precipitation of metal sulphides; and

- Other methods such as: adsorption and elution, and disposal of the concentrated thiosalts solution, reduction by metal (*e.g.* Fe) and sea disposal.

Many of above methods were found to be technically feasible, but expensive. Climate affects the performance of biological processes. The "natural degradation" in tailings ponds providing several days to weeks exposure to air and sunlight is the most common method. Organic compounds such as frothers and collectors contained in process effluents could also be degraded (Kuyucak *et al.*, 2001).

Some sites practice the "increased buffering capacity" method in the treated waters by first increasing the pH to levels >10 to remove metal ions (*e.g.* Zn), and then decreasing the pH in decant to levels <9 by adding carbon dioxide (CO_2) in order to meet regulatory discharge limits. This method may function satisfactorily for thiosalts concentrations <100 mg/l (Kuyucak *et al.*, 2001). Similarly, the use of sodium bicarbonate ($NaHCO_3$) was also tested to increase buffering capacity in treated waters (Li and Boucher, 1999; Li, 2004).

Hydrogen peroxide is currently being used for treating tailings reclaim water at several mining sites (*e.g.* Boliden's Apirsa in Spain, X-Strata Zinc New Brunswick in Canada, X-Strata Kidd Metallurgical site in Canada), before a lime neutralization/precipitation (HDS) process (Kuyucak and Yaschyshyn, 2007). The biological oxidation methods were examined with pilot-scale tests on site in Canada. Biological reduction of thiosalts to elemental sulphur (S^0) is investigated with laboratory tests as an alternative method. For low concentrations of thiosalts (*e.g.* <100 mg/l as S_2O_3), the use of $NaHCO_3$ or increased pH levels (*e.g.* pH >10 to remove metals and other ions) along with CO_2 pH regulation are also practiced.

Biological Sulphate Precipitation and Passive Processes

The main objective of passive water treatment systems is to accelerate naturally occurring reactions in a confined system rather than in a receiving natural water body, such as lakes or streams (Hedin *et al.*, 1994). They have, in some circumstances, been proven to be feasible alternative to conventional lime neutralization/precipitation methods. Characteristics of drain water from mines can change significantly and are usually unique to each site. Depending on the water characteristics, several passive treatment systems can be used, such as aerobic, anaerobic, limestone, or their combinations.

Aerobic Systems

In aerobic systems, metals such as iron (Fe^{2+}, Fe^{3+}), manganese and aluminum can be effectively removed by oxidation and hydrolysis reactions (Faulkner and

Skousen, 1994). In oxidizing environments, adequate aeration, the presence of alkalinity, and retention time are the primary design parameters for mine drainage having a neutral pH and containing iron and/or manganese. Oxygen by aeration stimulates oxidative reactions, while retention time provides the required time for reactions to occur and precipitation to take place. The abiotic iron oxidation reaction is faster in alkaline or neutral waters than in acidic waters. Rapid hydrolysis can be ensured at pH levels above 3. The reactions for the removal of Fe are shown in Equation (13) and (14):

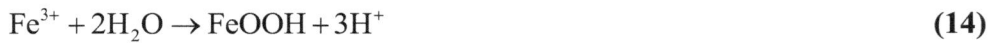

$$Fe^{2+} + \frac{1}{4}O_2 + H^+ \rightarrow Fe^{3+} + \frac{1}{2}H_2O \tag{13}$$

$$Fe^{3+} + 2H_2O \rightarrow FeOOH + 3H^+ \tag{14}$$

Ironhydroxide precipitates, such as FeOOH, that are generated in passive systems, are usually less fluffy and voluminous than iron hydroxide precipitates (*e.g.* $Fe(OH)_3$) produced by the straight lime neutralization process. As a result, these precipitates do not occupy a large space. An adequately designed system can be operated for a long time (*e.g.* >20 years). The area requirement for the treatment is determined based on the total loading of acidity and/or Fe. The surface of FeOOH precipitates provides adsorption sites for other metals (*e.g.* Zn, Cu, Mn, *etc.*) which results in removal of these metals in addition to other removal mechanisms (*e.g.* neutralization/precipitation, carbonation, *etc.*). Example reactions for other metals undergoing hydrolysis and removal from the water in aerobic conditions are shown in Equation (15) and (16) (Gazea *et al.*, 1996):

$$Al^{3+} + 3H_2O \rightarrow Al(OH)_3 + 3H^+ \tag{15}$$

$$Mn^{2+} \frac{1}{4}O_2 + \frac{3}{2}H_2O \rightarrow MnOOH + 2H^+ \tag{16}$$

It should be noted that hydrolysis and precipitation reactions in Equations (15) and (16) produce proton (H^+) acidity, which is also called mineral acidity. For a successful process, the system must contain sufficient alkalinity to neutralize the produced acidity.

Anaerobic Systems

Aerobic-organic substrate systems with the presence of sulphate reducing bacteria (SRB) provide conditions for removal of heavy metals such as Cu, Ni, Zn, Cd, and Pb from mine waters. Conversion of sulphate (SO_4) to sulphide (S^{2-}) in ARD by SRB can generate bicarbonate alkalinity as shown in Equation (17) (Kuyucak *et al.*, 2006). The produced S^{2-} removes metals forming insoluble complexes with them, as shown in Equation (18). These systems are often called anaerobic wetlands or subsurface flow wetlands.

$$2CH_2O + SO_4^{2-} \rightarrow H_2S + 2HCO_3^-$$
(17)

$$M^{2+}H_2S + 2HCO_3^- \rightarrow MS + 2H_2O + 2CO_2$$
(18)

CH_2O and M refer respectively to organic substrate (nutrient) and to metal.

These systems are more complicated and expensive than aerated systems. However, removal of heavy metals from ARD and production of alkalinity, resulting in an increase in pH, can only be achieved by these systems. The process kinetics depend on several parameters, such as temperature, pH, concentration of SO_4, redox potential, nutrient availability, loading of metals, and flow rates.

Limestone Drains

Limestone drains (LD) are used to neutralize the total acidity and to increase alkalinity and pH. The proton acidity found in ARD dissolves calcium carbonate and produces bicarbonate alkalinity. The reaction is illustrated in Equation (19).

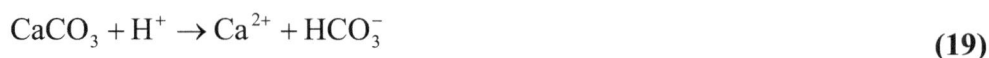

$$CaCO_3 + H^+ \rightarrow Ca^{2+} + HCO_3^-$$
(19)

The presence of high concentrations of Al, Fe^{3+}, and SO_4 (*e.g.* >2,000 mg/l of SO_4) is undesirable for LD as their precipitates can clog the system. LDs can either be used as a pretreatment prior to aerobic or anaerobic wetland treatment or be placed after a wetland to modify the pH in the final effluent.

The most common passive treatment systems are sulphate reducing bacteria based processes, anoxic limestone drains, constructed anaerobic and aerobic wetlands, and biosorption. Experience has demonstrated that the influent flow rate, contaminant concentrations, pH and alkalinity (or acidity) are all important elements to be considered for the system performance. In addition, the capacity of a biological treatment system is significantly affected by ambient temperatures and changes in the pH.

Sulphate Reducing Bacteria "SRB" Based Processes

Under reducing and anaerobic conditions, and in the presence of organic carbon nutrient sources, sulphate reducing bacteria (SRB) convert AMD sulphate to sulphide, as illustrated by Equations (20) and (21).

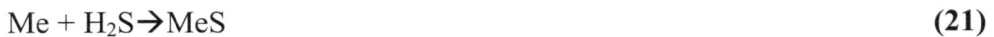

$$2CH_2O + SO_4^{2-} \rightarrow HS^- + 2\ HCO_3^- + H^+ \tag{20}$$

$$Me + H_2S \rightarrow MeS \tag{21}$$

The use of SRB in engineered lagoons, in open pits and flooded mines, has been investigated worldwide and have received greater attention (Kuyucak and St-Germain, 1994; Kuyucak 2002, Kuyucak and Palkovits, 2009). Their use in a controlled reactor was implemented by Budelco (in The Netherlands) at full scale to remove metals and SO_4 from underground mine water in the early 1990s. Recently several new developments are underway to be either pilot tested or implemented at mining sites to treat AMD and recover valuable metals for sale or use (Bratty *et al.*, 2006). Sometimes, limestone and soil are also added into the nutrient mixture to increase alkalinity and, subsequently, enhance the activity of the SRB.

Alkalinity can be generated by the dissolution of limestone or other carbonates rocks; bacterial sulphate reduction generates sulphite to precipitate/remove metals, too.

Base metal ions including Pb^{2+}, Zn^{2+}, Fe^{2+} *etc.* in net-acidic AMD can be removed with the help of anaerobic sulphide reducing bacteria. Removal of Al occurs by precipitation which is dependent on pH and alkalinity levels in the treatment media. Ferric iron (Fe^{3+}) can undergo hydrolysis and iron hydroxide precipitation occurs in solutions where pH levels are >4.5. In passive treatment systems, reduction in manganese levels is usually marginal (Kuyucak, 2002; Kuyucak *et al.*, 2006).

Implementation of passive treatment systems to a decommissioned mining site in Val d'Or in Quebec, Canada, in 2004, proved that passive systems can be a cost-effective alternative to active systems (Kuyucak *et al.*, 2006; Kuyucak and Chabot, 2009). The quality of the produced water is in compliance with the provincial government regulations. The site-specific passive treatment facility included: a seepage collection system; anaerobic and aerobic cells; and a limestone filter. Critical parameters for successful operation of the system included: nutrients and organic substrate biodegradation; anaerobic/reducing conditions; hydraulic loading and metal loading rate changes; storm water impact; hydraulic design of the system (*i.e.* to avoid short-circuiting and channelling); gas lock-up; and temperature (Kuyucak, 2006). It has been revealed that passive systems should be started at relatively high ambient temperatures for optimal performance.

The SRB-hosting organic substrate should be protected against freezing conditions by covering it with a thick soil layer (*i.e.* >0.6 m). Burying the organic substrate containing the bacteria to conserve heat and prevent freezing was an approach used by several system designers and operators (Kuyucak *et al.,* 2006).

Wetlands

A wetland can remove organic and inorganic compounds and suspended solids (Kadlec and Knight, 1996). Wetlands can be designed as surface flow (aerated) or subsurface flow (anaerobic) wetlands and can be vegetated with aquatic plants or operated without plants. In a wetland system water can flow horizontally or vertically. The design of a vertical flow wetland system is similar to the design of granular media wetlands where sand and gravel are often used as granular media. It can also be called as wetland biofilter (Rozema, 2010). A wetland is usually composed of an Oxidation Zone, which contains aquatic plants, and a Reducing Zone, which is the sedimentation zone rich in SRB, denitrifying and Mn reducing bacteria. Plants play a filter role, taking up metals and helping oxidation processes to occur, while bacteria act as catalysers for chemical reactions. The type of wetland, aerobic or anaerobic or combined, is decided based on the quality of wastewater. Aerobic wetlands are suitable for waters containing net alkalinity. Anaerobic wetlands are suitable for acid waters containing metal ions. Designing wetland treatment systems for waters containing Al and Fe in high concentrations (*e.g.* >10-15 mg/l) requires special considerations to accommodate hydrolysis and precipitation of hydroxide complexes of the metal ions.

Anoxic Limestone Drains (ALD)

Two main objectives are generally considered in designing an ALD to obtain optimum acid and metal ion removal: (1) generate sufficient alkalinity to neutralize the acidity, and (2) decrease the metals loading through oxidation/hydrolysis and precipitation mechanisms. The discharge water from ALD should have a pH close to neutrality and low metal concentrations (Maqsoud, *et al.*, 2007). An ALD system generally consists of an excavated seepage interception trench backfilled with crushed limestone and covered with plastic and clay-soil to keep air out. As acid water flows through the system, it dissolves limestone (Warkentien, 2010) causing an increase in alkalinity, oxidation of Fe^{2+} and formation of ferric oxyhydroxides. Designs of ALD are reported to be site-specific taking into account the water quality, climate, site topography, space *etc.* Usually, an ALD system is followed by a wetland for oxidation, precipitation and separation of iron and other contaminants.

Open Limestone Trenches

Trenches are typically most effective for polishing/conditioning AMD with relatively low strength and acidity. A trench contains a layer of limestone where alkalinity and buffering capacity in the acid water are increased. A process consisting of aeration, mixing and open limestone trenches that was designed and operated at a decommissioned Barrick Gold site in Val d'Or, Quebec, has been operated successfully year round since 1997. The system generates an effluent quality that meets Menviq standards, especially for parameters including pH, Ni, Cu and Fe (Kuyucak and Chabot, 2009).

Biosorbents and Peat Filters

A bed of biosorbents such as sawdust, sphagnum moss or algae can be placed where the seepage occurs. When it is saturated by metal ions, the biosorbent materials can either be disposed of with tailings or recycled to a smelter, or washed with an appropriate eluate for recovery of metals (Kuyucak, 1990;

Kuyucak and Volesky 1990; Kuyucak 2002; Volesky and Kuyucak, 1988; Gusek, *et al.*, 2006).

The peat consisting of sphagnum moss has been reported to be functional for removal of boron from wastewaters under alkaline conditions (Sartaj *et al.*, 2000). In addition, peat has shown ability to remove BOD_5/COD, TSS, NH_4-N and metal ions (*e.g.* Zn, Fe, Cu) at wide pH range (Brooks, 2001; Kuyucak and Zimmer, 2004) from various municipal and industrial wastewaters. In its dehydrated form, peat is a highly effective absorbent for water. Application of peat filters for separation of contaminants from mining effluents is possible. A peat filter made of air-dried and compacted under very controlled conditions to obtain a design hydraulic conductivity of about 10^{-4} cm/s can be used for separation of metal ions and NH_3/NO_3 from effluents.

CONCLUSIONS

Industrial wastewaters such as mining effluents resulting from mining and metallurgical activities and landfill leachate stemming from accumulation and storage of industrial and/or municipal wastes may require treatment for separation of a variety of contaminants. These contaminants may include metal ions and metalloids, ammonium and nitrate, phosphate, thiosalts, cyanide and acids. Although treatment requirements for a given site are defined on the basis of *on site* specific discharge regulations and wastewater characteristics, generally primarily application of BATEA is considered. Sustainability considerations as to the use/reuse or recycling of the treated water and management of sludge produced by the treatment process should be also included in the selection and design of an appropriate treatment process. In addition to the effectiveness of the treatment method to remove the required contaminants, operational and capital costs including operational preference (*i.e.* level of automation), energy and chemical usage, land availability and site conditions (*e.g.* temperature, weather, altitude, *etc.*) play a key role in selecting the desired treatment system. Careful assessment of the overall site water balance and quality and quantity of water sources can help in optimizing the objectives and the size of the treatment system. Based on the sources, typology and concentrations of contaminants that need to comply with regulatory standards, the use of more than one treatment system can also be considered for a specific site. For instance, mine dewatering waters and tailings reclaim waters can be managed separately. Innovative approaches, such as addition of streams at different points along the system and recycling of sludge, can be incorporated in the design and adopted in the operation of the treatment systems to increase their efficiency and reduce the overall cost by decreasing the chemical and/or energy usage.

ACKNOWLEDGEMENT

Declared none.

CONFLICT OF INTEREST

The author(s) confirm that this chapter content has no conflict of interest.

REFERENCES

Allen, E.W. (2008). Process water treatment in Canada's oil sands industry: II A review of emerging technologies. Report, *Journal of Environmental Engineering and Science*, 7(2), 123-138.

Bratty, M., Lawrence, R., Kratochvil, D., Marchant, B. (2006). Applications of Biological H_2S Production from Elemental Sulphur In The Treatment of Heavy Metal Pollution Including Acid Rock Drainage, In: *Proceedings of 7th ICARD "Leadership: Gateway to the Future"*, St. Louis, Missouri, USA.

Bioteq (2010). *Sulf-IX Pilot*. Available from: http://bioteq.ca/water-treatment/operations/sulf-ix-pilot/ [Accessed 17 February 2012].

Brooks, J.L. (2001). *Peat Filter Septic Systems*. Available from: http://www.barnstablecountyhealth.org/ia-systems/information-center/compendium-of-information-on-alternative-onsite-septic-system-technology/peat-filter-septic-systems [Accessed 17 February 2012].

Environment Canada (2001). *Metal Mining Requirements*, Environment Canada Publications. Available from: www.ec.gc.ca/eem [Accessed 17 February 2012].

Faulkner, B.B., Skousen, J.F (1994). *Treatment of acid mine drainage by passive systems*.P.250_257. US Bur. Mines SP-06A-94.

Gazea, B., Adam, K., Kontopoulos, A. (1996). A review of passive systems for the treatment of acid mine drainage. *Minerals Engineering*, 9, 23-42.

Gunther, P., Mey, W., van Niekerk, A.M (2006). Water reclamation and the drive to sustainability in mining, Emalahleni, South Africa – Case Study, In: *Proceedings of Water in Mining 2006 Conference*, Anglo Coal South Africa.

Gusek, J.J., Wildeman, T.R., Conroy, K.W. (2006). Conceptual methods for recovering metal resources from passive treatment systems, In: *Proceedings of 7th ICARD "Leadership: Gateway to the Future"*, St. Louis, Missouri, USA.

Hedin, R.S., Nairn, R.W., Kleinmann, R.L.P. (1994). *Passive treatment of coal mine drainage.* Bureau of Mines IC 9389. Washington, DC.

Hedin, R.S. (2006). Recovery of a marketable iron product from coal mine drainage, In: *Proceedings of 7th ICARD "Leadership: Gateway to the Future"*, St. Louis, Missouri, USA.

INAP (2003). *Treatment of Sulphate in Mine Effluents.* Available from: http://www.inap.com.au/public_downloads/Research_Projects/Treatment_of_Sulphate_in_Mine_Effluents_-_Lorax_Report.pdf [Accessed 17 February 2012].

Kadlec, R.H., Knight, R.L. (1996). *Treatment Wetlands.* CRC Press LLC.

Kuyucak, N. (1990). Feasibility of biosorbent applications. In: *Biosorption of Heavy Metals,* Ed. B. Volesky, CRC Press.

Kuyucak, N., Volesky, B. (1990). Biosorption by algal biomass. In: *Biosorption of Heavy Metals,* Ed. B. Volesky, CRC Press.

Kuyucak, N., Sheremata, T.M., Wheeland, K.G. (1991). Evaluation of improved lime neutralization processes, part I: lime sludge generation and stability, In *Proceedings of 2nd Int. Conf. on the Abatement of Acidic Drainage*, Montreal, Canada, *2*, 1-14.

Kuyucak, N., Mikula, R.J., Wheeland, K.G. (1991). Evaluation of improved lime neutralization processes, part I - III: Interpretation of properties of lime sludge generated by different processes, In: *Proceedings of 2nd Int. Conf. on the Abatement of Acidic Drainage*, Montreal, Canada.

Kuyucak, N., Rao, R., Sheremata, T., Finch, J. (1994). Value recovery from acid mine drainage. *MEND/CANMET Report, MEND Proj.* No. 3.21.2a.

Kuyucak, N., St-Germain, P. (1994). *In situ* treatment of acid mine drainage by sulphate reducing bacteria in open pits: scale-up experiences, In: *Proceedings of Int. Land Reclamation and Mine Drainage Conf. and 3rd Int. Conf. on Abatement of Acid Drainage*, Pittsburgh, PA.

Kuyucak, N. (1995). Discussion: effect of pH on metal solubilization from sewage sludge. *Canadian Journal of Civil Engineering.*

Kuyucak, N., Payant, S. (1995). Lime neutralisation methods for improving sludge density and final effluent quality, In: *Proceedings of CIM, 2nd Int. Symp. on Waste Processing and Recycling*, Vancouver, BC.

Kuyucak, N., Payant, S., Sheremata, T. (1995). An improved lime neutralization process, In: *Proceedings of Sudbury '95 Mine Environment Conf.*, Sudbury, Ontario.

Kuyucak, N., Sheremata, T. (1995). *Lime neutralisation process for treating acid waters.* U.S. Patent: 5.427.691.

Kuyucak, N. (1998). Mining, the environment and the treatment of mine effluents. *International Journal of Environment and Pollution, 10,* 2/3.

Kuyucak, N., Lindvall, M., Serrano, J.A.R., Oliva, A.F. (1999). Implementation of a high density sludge "HDS" treatment process at the Boliden Apirsa mine site, In: *Proceedings of Clean Technologies Wastewater Treatment Symposium*, Sevilla, Spain.

Kuyucak, N. (2000). Microorganism, biotechnology and acid rock drainage. *Int'l Minerals and Metallurgical Processing, 17,* 85-96.

Kuyucak, N. (2001a). AMD Prevention and Control Options. *Mining Environment Management Journal*, January, 12-15.

Kuyucak, N. (2001b). Acid mine drainage (AMD) - Treatment options for mining effluents. *Mining Environment Management Journal*, 14-17.

Kuyucak, N. (2001c). Thiosalts in mill effluents, related environmental issues and treatment options, In: *Proceedings of 4th Int'l Symposium on Waste Processing and Recycling in Mineral and Metallurgical Industries, (CIM Conference)*, Toronto ON, Canada.

Kuyucak, N., Serrano, J.R.A., Hultqvist, J., Eriksson, N. (2001). Removal of thiosalts from mill effluents "studies conducted at the Boliden Apirsa Site", In: *Securing the Future 2001, Mining and the Environment Conference proceedings.*

Kuyucak, N., Lindvall, M., Sundqvist, T., Sturk, H. (2001b). Implementation of a high density sludge "HDS" treatment process at the Kristineberg mine site. Securing the Future 2001, In: *Mining and the Environment Conference proceedings.*

Kuyucak, N. (2002). Microorganisms in mining: generation mitigation and treatment, *EJMP & EP*, 2/3.

Kuyucak, N., Chávez, J., del Castillo, J.R., Ruiz, J. (2003). Technical feasibility studies and uses of treated AMD at Kingsmill Tunnel, Peru, In: *ICARD'03 Conference Proceedings*, Cairns, Australia.

Kuyucak, N., Chávez, J., del Castillo, J.R., Ruiz, J. (2004). Potential use of treated acid mine drainage at Kingsmill Tunnel, Peru - Technical and economic feasibility studies, In: *5th Int'l Symposium on Waste Processing* and *Recycling in Mineral and Metallurgical Industries, (CIM Conference)*, Hamilton ON, Canada.

Kuyucak, N., Zimmer, M. (2004). Natural systems successfully treating landfill leachate, In: *ISWA 2004 Conference Proceedings*, Rome, Italy.

Kuyucak, N. (2006). Selecting suitable methods for treating mining effluents, In: *Water in Mining – 2006: Multiple Values of Water Conference*, Brisbane, Australia.

Kuyucak, N., Chabot F., Martschuk, J. (2006). Successful implementation and operation of a passive treatment system in extremely cold northern Quebec, Canada, In: *Proceedings of 7th ICARD "Leadership: Gateway to the Future*, St. Louis, Missouri, USA.

Kuyucak, N. (2007). Sources of mining effluents and suitable treatment options. In: *Proceedings of International Mineral Processors Conference (IMPC)*, Ankara, Turkey.

Kuyucak, N., Yaschyshyn, D, (2007). Managing thiosalts in mill effluents "Studies conducted at the Kidd metallurgical site", In: *Sudbury 2007, Mining and Environment Conference*, Sudbury, Ontario, Canada.

Kuyucak, N, (2008). Thiosalts in mill effluents and possible management options, In: *Proceedings 11th International Mineral Processing Symposium*, Bellek, Antalya, Turkey.

Kuyucak, N., Chabot, F, (2009). Successful operation of passive water treatment systems in extremely cold Canadian climate, In: *Proceedings of Securing the Future and 8th ICARD*, Skelleftea, Sweden.

Kuyucak, N., Palkovits, F. (2009). Integrated approach for water management in mining, In: *Proceedings of The 2009 Paste -12th International Seminar on Paste and Thickened Tailings*, Vina del Mar, Chile.

Li, M., Boucher, J.F. (1999). Thiosalts treatment by bicarbonate addition - A laboratory-scale feasibility study, *NTC Project 41341-60, Report submitted to BMD and Thiosalts Consortium, May, 131.*

Li, M. (2004). *Cost Estimates for top ranking thiosalts Treatment options at the Kidd Metsite* (Internal Noranda/Falconbridge Memo to David Yaschyshyn, dated January 13, 2004).

Maqsoud, A., Bussière, B., Aubertinl, M., Potvin, R., Cyr, J. (2007). Evaluation of hydraulic residence time in the limestone drains of the Lorraine site, Latulippe, Québec, In: *Proceedings of Mining and the Environment IV Conference*, Sudbury, Ontario, Canada.

MEND Report (1994). *Status of Chemical Treatment and Sludge Management.* CANMET, NRC and Canada.

Metcalf and Eddy Inc. (1991). *Wastewater Engineering: Treatment, Disposal and Reuse.* Third Edition. McGraw-Hill, Inc.

Nelson, G.M., Kroeger, E.B., Arps, P.J. (1998). Chemical and Biological Destruction of Cyanide: Comparative Costs in a Cold Climate. *Min. Pro. Ext. Met., 19, 217-226.* Gordon and Breach Publishers. Eds: F.M. Doyle, N. Arbiter and N. Kuyucak.

NRCan (2010). *Cyanide Recycling Technologies.* Available from: http://www.nrcan-rncan.gc.ca/mms-smm/tect-tech/ser-ser/met-tem-eng.htm#Cyanide [Accessed 17 February 2012].

Ontario Drinking Water Standards (2000). *Regulation 459/00,* August 2000.

Randol. (1998). Mining and Milling Processes: Treatment of Mine/Mill Effluents, 2954-55.

Rao, R., Kuyucak, N., Sheremata, T., Finch, J. (1994). Prospect of recovery / recycle from acid mine drainage., In: *Int. Land Reclamation and Mine Drainage Conf.* and *3rd Int. Conf. on Abatement of Acid Drainage,* Pittsburgh, PA.

Ripley, E.A., Redmann, R.E., Crowder, A.A. (1995). *Environmental Effects of Mining.* St-Lucie Press. Pp. 181-197.

Ritcey, G.M. (1989). Tailings Management: Problems and Solutions in the Mining Industry. *Process Metallurgy 6, Elsevier Publication,* 604-650.

Rozema, R.L. (2010). *Natural Treatment Systems.* Available from: http://www.aqua-tt.com/media/rozemacv.pdf [Accessed 17 February 2012].

Sartaj, M., Fernandes, L., Castonguay, N. (2000). Removal of boron and other contaminants from landfill leachate by peat and the effect of lime addition, In: *Proceedings of16th Eastern Region Conference Canadian Association on Water Quality.*

Smith, A., Mudder, T.I. (1991). *The Chemistry and Treatment of Cyanidation Wastes.* Mining Journal Books Publishers, London.

Smith, A., Mudder, T.I. (1994). An Environmental Perspective on Cyanide. *Mining World News,* 6/9.

Smith, J.P. (2000). Potable Water from Sulphate Polluted Mine Sources. *Mining Environmental Management Journal,* 8, 7-9.

van Niekerk, A.M., Wurster, A., Boase, A., Cohen, D. (2006). Water reclamation and the drive to sustainability in mining; Emalahleni, South Africa – Case study, In: *Water in Mining 2006 Conference,* Anglo Coal South Africa, Clewer 1036, South Africa.

Vigneault, B., Holdner, J., Bélanger, J. (2003). Validation of an anion exchange method for the preservation and analysis of thiosalt speciation in mining waste waters. *CANMET, Project MMSLNo. 602507.*

Volesky, B., Kuyucak, N. (1988). *Biosorbent for gold.* U.S. Patent: 4.769.223.

Warkentien, A. (2010). *What Is an Anoxic Limestone Drain?* Available from: http://www.ehow.com/facts_6887507_anoxic-limestone-drain_.html [Accessed 17 February 2012].

World Bank (1997). Pollution Prevention and Abatement Handbook.

Yanful, E.K., Wheeland, K.G., St-Arnaud, L., Kuyucak, N. (1991). Overview of Noranda research on prevention and control of acid mine drainage. *AMIC Workshop,* Perth, Australia.

Zinck, J.M., Griffith, W.F. (2000). An assessment of HDS-type treatment processes efficiency and environmental impact, In: *ICARD'2000,* II, 1027-1035.

Send Orders of Reprints at reprints@benthamscience.org

CHAPTER 10

Evaluation of Sediments' Contamination by Hyperspectral Analysis

V. Gente[1,*], S. Geraldini[2], F. La Marca[3] and F. Palombo[4]

[1]Environmental Engineer, Italy; [2]Istituto Superiore per la Protezione e Ricerca Ambientale, Rome, Italy; [3]Department of Chemical Engineering Materials and Environment, University of Rome "La Sapienza", Rome, Italy and [4]Golder Associates S.r.l., Rome, Italy

Abstract: Sediments that are dredged during routine operations for maintenance of harbour areas are generally contaminated, even in consideration of given standards or regulations that do not allow their free disposal in the aquatic system. Therefore, sediments coming from dredging operations have to be properly characterised since their final destination depends on the level of contamination.

Due to the quantity of dredged sediments and the number of contaminants to determine, analyses carried out for characterisation are expensive and time-consuming.

In this research work, a new approach for a prompt evaluation of sediments' contamination based on hyperspectral analysis is proposed. Results of laboratory tests carried out on marine sediments, before and after chemical-physical treatments, show a correlation among the value of reflectance, obtained by hyperspectral analysis, and the level of contamination and size distribution of sediments.

Keywords: Contaminated sediments, Heavy metals, Hyperspectral analysis.

INTRODUCTION

Dredging of sediment represents an issue of great social and economic relevance in Italy, due to the large coastal development (about 7,500 km of coastline).

Port traffic represents an important economic resource and, in order to be competitive, several ports need to realize infrastructural works and to deepen navigational channels for port development. Furthermore, dredging for ordinary maintenance takes place periodically in order to preserve navigability.

*Address correspondence to V. Gente: Environmental Engineer, Italy; Tel:+39069699532; Fax: +39069699532; E-mail: vincenzo.gente@ingpec.eu

In most cases, the big port areas serve large industrial sites, where extremely heterogeneous and often heavily impacting activities are located. Coastal areas are the receptors of mineral or organic solid particles, coming from natural chemical and physical processes, and of contaminants, discharged by industrial effluents into water bodies, sea, river or lake, that eventually build up in sediments (Jones *et al.* 1998; Olin-Estes and Palermo, 2001). Moreover, most of the Italian ports are included within the so called "National Relevance Contaminated Sites". These sites, distributed all along the Italian coast, are heavily impacted sites for which the Italian Government has allocated specific resources for remediation: they are extremely heterogeneous for extension, geo-morphological characteristics, hydrodynamics, contamination history and uses.

In Italy dredging activities (for maintenance and deepening of port access channels, building of port facilities, remediation of contaminated areas or laying of cables and pipelines) produce huge quantities of seabed material to be managed, with different characteristics in terms of grain size and contamination.

The common approach of the Italian legislation for the management of dredging processes is based on a preliminary characterization phase, to determine physical, chemical and microbiological characteristics of the entire volume of sediments to be dredged and to select the different management options on the basis of the sediment quality (Okx and Stein, 2000, Rivett *et al.*, 2002).

Characterization activities require a lot of time and money, both for sampling and analysis phase. For this reason, in this work, a new approach for a prompt evaluation of sediments' contamination based on hyperspectral analysis is proposed.

Techniques based on the processing and analysis of digital images are used in many industrial sectors, such as mining and recycling industries (Cronhjort *et al.*, 1997). Digital imaging based approaches can be profitably adopted in order to obtain specific information that allows the recognition and classification of, *inter alia*, objects, materials, and chemical species. The differences in their exploitation depend on the typology of images acquired and of extracted features, which can result mainly in the analysis of morphological, textural or spectral characteristics. Therefore, digital image based techniques can be applied for many purposes, such as beneficiation of raw materials, sorting and separation processes, quality monitoring, and material characterisation. The main advantages of this kind of techniques can be found in the possibility of obtaining a prompt and reliable response and in the adoption of automated processes (Bonifazi *et al.*, 2002).

The aim of this research is to evaluate the reliability of a specific technique based on the acquisition and analysis of hyperspectral images (imaging spectrophotometric analysis) for a fast classification of chemical-physical characteristics of sediments.

MATERIALS AND METHODS

For the present research work, sediments have been collected during dredging operations in the area of Bagnoli, that is located near Naples, in the center of Italy, and is included in the 26 marine and brackish sites classified as "National Relevance Contaminated Sites". After dredging, sediments have been preserved in plastic bins. In order to obtain the proper amount of sediments, samples of dredged material have been collected after homogenizing.

In order to assess the feasibility of technique to evaluate the contamination of sediments, chemical-physical treatments have been carried out on samples of dredged sediments. In this way, two fractions have been obtained, a contaminated and a clean fraction, with different chemical-physical characteristics, notably with different contamination levels and size distributions. The products of the separation processes, in terms of contaminants' content and size distribution, have been evaluated both by mean of chemical-physical analyses and by the application of imaging hyperspectral analysis.

Chemical-Physical Treatments

Laboratory tests have been carried out adopting three different technologies. In particular the collected sediments have been submitted to sieving, hydrocycloning and flotation tests. These treatments have been selected since they can be carried out on wet sediments and, therefore, they can be realized as on-site treatments nearby coastal areas or on mobile-platforms, without preliminary dewatering operations.

The operating conditions of the laboratory tests are summarized in the Table **1**. Detailed descriptions of the devices and conditions adopted for the treatments of dredged sediments can be found in Gente *et al.*, 2009.

Table 1: Operating conditions of the laboratory tests

Sieving Test	Hydrocyclone Test	Flotation Tests	
		Test A	*Test B*
Wet sediments	Solid-liquid weight ratio: 1:15	Solid-liquid weight ratio: 1:3	Solid-liquid weight ratio: 1:3
Sieves of stainless steel woven wires	50 mm hydrocyclone Mozley C700 MK II	3 l Denver Cell	3 l Denver Cell
2.00 mm, 125 μm and 63 μm	Vortex finder: 14 mm spigot cap: 6.4 mm	Chemical reagent: pine oil	Chemical reagents: pine oil+amylxanthate

Hyperspectral Analysis

Hyperspectral analysis, based on imaging spectrophotometry, has been carried out adopting an integrated hardware and software architecture. As a matter of fact, imaging spectrophotometry consists in the capture and analysis of *spectra* that are obtained by acquiring and processing digital image sequences of the surface of properly energized samples of materials. According to the different wavelength of the source and to the different device spectral sensitivity, different physical-chemical superficial characteristics of the samples can thus be investigated and analysed (Spectral Imaging Ltd, 2002).

The result of the analysis is a spectrophotometric measure, represented by a plot of the value of reflectance obtained for the different wavelengths investigated.

The imaging based spectrophotometry architecture set-up, adopted in this study, is based on the utilization of the *ImSpector*™ V10 spectrometer device, from *SpecIm*™ Oy, working in the spectral range from 400 to 1000 nm ± 5 nm.

The spectrophotometer, *ImSpector*™ V10 (Fig. **1**) is constituted by optics based on volume type holographic transmission grating. The grating is used in patented prism-grating-prism construction (PGP element), which can be shown to have high diffraction efficiency, good spectral linearity and it is nearly free of geometrical aberrations due to the on-axis operation principle. A collimated light beam is dispersed at the PGP so that the central wavelength passes symmetrically through the grating and prisms (so that it stays at the optical axis) and the short and longer wavelengths are dispersed up and down compared to central wavelength.

The spectrum is measured thanks to an array detector constituted by a CCD (Charged Coupled Device) monochrome camera. The availability of a detector constituted by a linear array of sensing elements, each one able to detect the spectral components of the corresponding investigated constituting domains of the object, could permit to measure the optical spectrum components and the spatial location of an object surface (Spectral Imaging Ltd, 2002). The resulting information is a digital image where each column represents the discrete spectrum values of the corresponding element of the sensitive linear array.

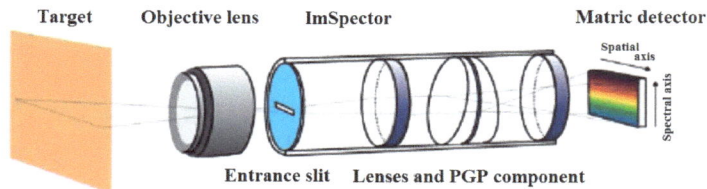

Figure 1: *ImSpector* operation.

The detection architecture system is basically constituted by four elements:

- A sample transportation plate (DV™ spectral scanner);

- A collimated illumination device (Fiber-lite) composed by one 150 W halogen lamp, as source light, and one illumination opening in optical fiber of 200 mm long and 2 mm width, positioned at 45° in respect to the transportation plate and presenting a minimum light divergence;

- The *ImSpector*™ V10 imaging spectrophotometer and coupled with a standard C-mount zoom lens; and

- An array detector, that is Teli CCD monochrome camera (Panasonic CS 8300).

The main technical characteristics of the *ImSpector*™V10 are the following:

- Spectral range: 400 – 1,000 nm;

- Image size: 6.6 (spectral) × 8.8 (spatial) mm;

- Lens and camera mount: Standard C-mount adapter.

Differently from other detection systems, always based on spectroscopy, such a system offers several advantages: i) reduced measurement time, ii) no scanning movement and iii) simultaneous measurement over a line area.

In many applications several single points can be replaced by one multi-channel imaging spectrometer thereby saving both in the instrumentation cost and in the amount of mechanics and required space. There are "disadvantages" too, namely the huge amount of data that should be collected and processed at once and stored for later processing.

Therefore, in some process application this may require sophisticated data processors, like DSP (Digital Signal Processors) circuits, or some kind of preliminary data reduction scheme before actual data processing. The latter one can be done electronically or mathematically using fast processors.

A preliminary step necessary to carry on, when an imaging spectrometry data handling and processing has to be applied is represented by calibration, that can be performed in three steps:

i)   Spectral axis calibration with spectral lamp(s);

ii) Dark image acquisition and storage; and

iii) Measurements and storage of "white reference image".

The operations described in the second and third points have to be repeated time to time, according to lighting and camera stability.

After the calibration phase: i) the image *spectra* is acquired, ii) the reflectance (R_{ci}) (at wavelengths i and spatial pixels c of interest) is computed:

$$R_{ci} = \frac{(\text{sample})_{ci} - (\text{dark})_{ci}}{(\text{white})_{ci} - (\text{dark})_{ci}} \qquad (1).$$

Such a procedure enables to compensate the offset due to CCD dark current and separates the sample reflectance from the system response.

In the present study, for each sediment sample, three measures have been carried out in order to assess the stability of the system.

Spectra Acquisition and Analysis

Spectra related to the different sample have been detected adopting an hyperspectral imaging based architecture able to assure a progressive and continuous horizontal translation of the different samples and the contemporary "synchronized" acquisition (at a pre-established step) of the *spectra* by the *ImSpector*™ V7 (Bonifazi *et al.*, 2005). A progressive series of scan lines was thus acquired. *Spectra* have been acquired from 400 nm to 700 nm, at 10 nm step, with high sensitivity. A full bi-dimensional spectral image, generated through the acquisition of each spectrum, associated to every investigated pixels (spatially identified along the scan line) was built.

Physical-Chemical Analysis

Physical and chemical analyses have been carried out in order to characterize dredged sediment before and after laboratory tests. In particular, physical analyses aimed at the evaluation of sediments size distributions, while chemical analysis have been carried out in order to determine the heavy metals' contents.

In order to determine size distribution, collected sediments and products coming from the different treatment tests have been analysed by manual sieving and, for particles with dimension <63 μm, by laser granulometer SYMPATEC HELOS/KA.

The results of the size-distribution analyses for the dredged sediment before treatment tests are shown in Fig. **2**. In terms of soil classification, the results showed an average content of 48% of sand and 52% of Silt.

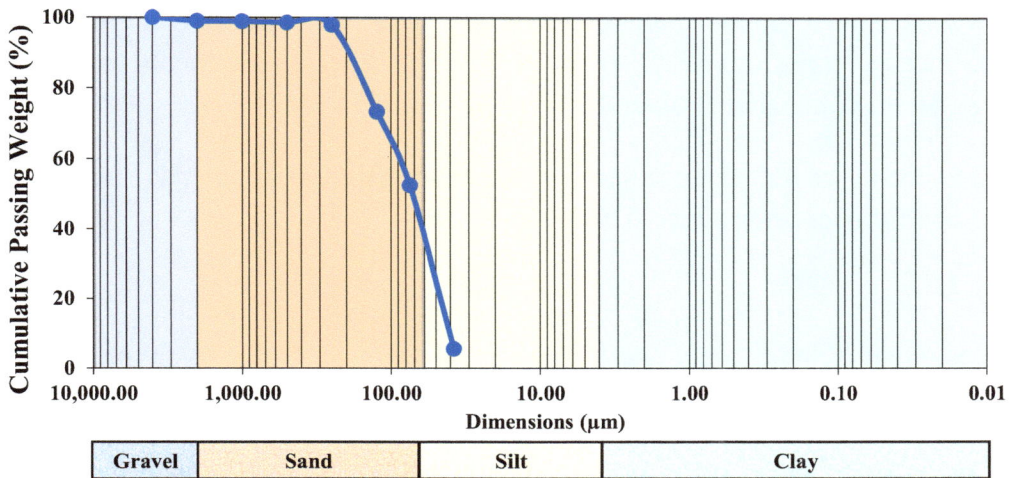

Figure 2: Size distribution and classification of dredged sediments.

In order to estimate the metals content, single mineralization has been carried out on sediment samples. The following metal species have been investigated: As, Ba, Cd, Cr, Fe, Mn, Ni, Pb, Cu and Zn.

The metal contents have been determined by Inductively Coupled Plasma-mass Spectrometry (ICP-OES), following the methodology EPA 3051A modified.

Table 2: Average metal content in dredged sediment

Metals	Concentration (ppm)
As	42.640
Ba	172.360
Cd	0.262
Cr	25.560
Fe	271700
Mn	2,023.663
Ni	11.560
Pb	215.226
Cu	25.797
Zn	523.870

The mineralization procedure, adopted to estimate the total metal concentrations on solids, has foreseen the microwave-assisted digestion of sediment using a Milestone mod. MLS1200 MEGA system. 0.5 g of sediment, after drying at 35 °C for 48 h have been grounded and digested with 6 ml of HNO_3 and 9 ml of HCl in an advanced composite vessel. After cooling, the digested sediments have been put in a 50 ml volumetric flask and diluted by distilled water.

The results of chemical analyses for the determination of metals content in collected sediments are shown in Table **2**.

RESULTS

Then sediments input and output from tests were analysed by hyperspectral methods.

Spectrophotometric data, carried out on the average of three measurements, show the significant part of the graph in the wavelength range between 600 nm and 700 nm, before and after this range the noise of the instrument is too high.

The results of chemical and of hyperspectral analyses are shown for each treatment typology carried out on the collected sediments.

Sieving Tests

According to the chemical analysis, the -63 μm size class, compared to the others, has the highest concentration values for chromium, copper, iron, manganese, nickel, lead and zinc. Arsenic, barium and cadmium concentrate in -2 mm +125 μm class.

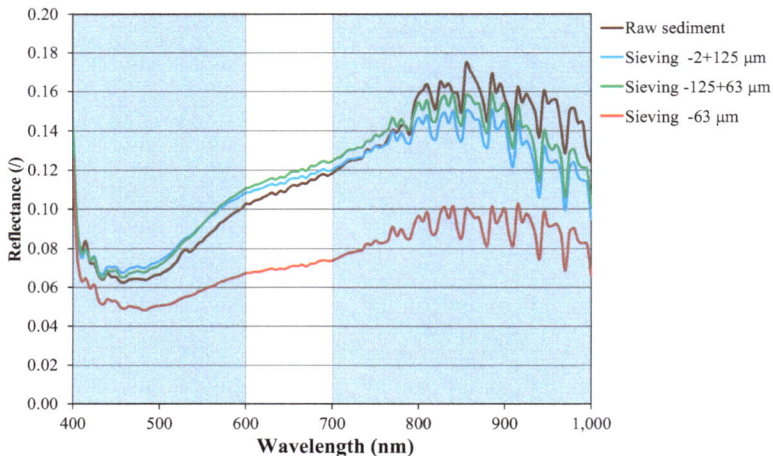

Figure 3: Sieving tests – *Spectra.*

Metal content of products of sieving tests are shown in Table **3**.

Fig. **3** shows spectrophotometric analysis results for sieving test. Spectral graph of fine fraction is at the bottom of the chart while the other *spectra* are at the top.

Table 3: Results of sieving tests

Size Class	-2 mm +125 µm	-125 µm +63 µm	-63 µm
Weight	13.3 (%)	48.5 (%)	38.2 (%)
Metals Content	ppm		
As	56.666	37.318	24.943
Ba	223.955	170.410	162.067
Cd	0.431	0.149	0.104
Cr	84.769	82.956	99.097
Fe	64,062.500	161,923.000	294,754.000
Mn	950.647	2,070.468	2,940.252
Hg	0.960	0.640	0.100
Ni	10.928	13.032	16.177
Pb	218.272	222.604	238.882
Cu	13.959	19.034	30.599
Zn	373.370	606.803	677.263

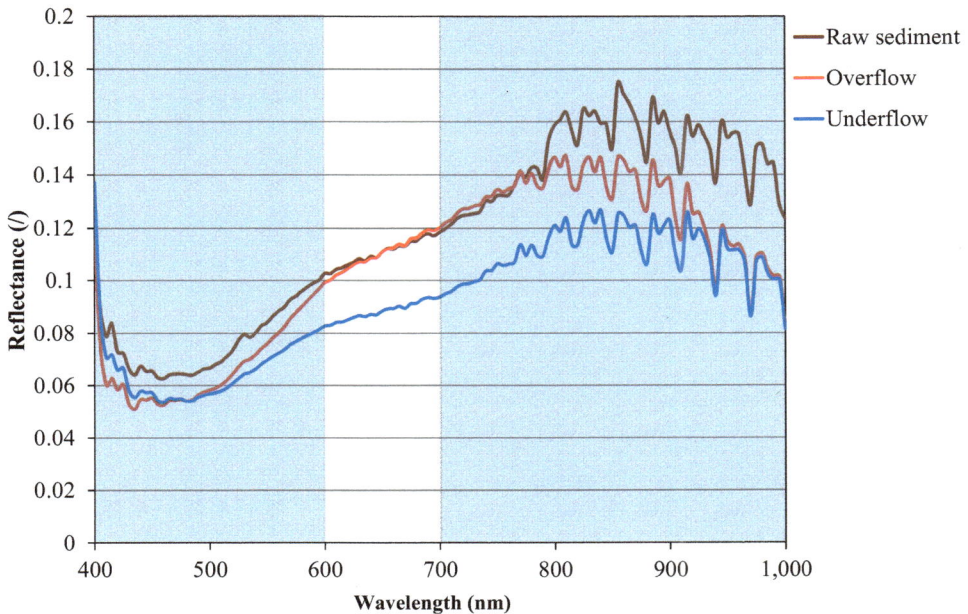

Figure 4: Hydrocyclone test – *Spectra.*

Hydrocycloning

According to chemical analysis, metals concentrate in underflow product.

Metal content for the overflow, underflow products and tailings, are shown in Table **4**.

In Fig. **4**, spectral graph of underflow is at the bottom of the chart while the overflow spectrum is at the top.

Table 4: Results of Hydrocyclone tests

Products	Overflow	Underflow
Weight	2.0 (%)	80.3 (%)
Metals	ppm	
As	33.077	96.276
Ba	161.902	209.089
Cd	5.711	25.609
Cr	55.549	316.916
Fe	149,325.000	296,464.000
Mn	2,151.108	3,704.724
Hg	1.320	2.500
Ni	14.712	21.989
Pb	191.959	907.551
Cu	21.519	63.130
Zn	602.832	1,743.870

Flotation Tests

According to chemical analysis, metal concentrations in float products, obtained in both flotation tests, are higher than metal concentrations in sink products, except for iron, manganese and barium.

Metal concentrations in flotation tests A and B are shown respectively in Table **5** and Table **6**.

Table 5: Results of flotation test A (Denver cell, reagent: pine oil)

Products (A)	Float	Sink
Weight	5.4 (%)	94.6 (%)
Metals	ppm	
As	42.616	45.282
Ba	97.377	162.399
Cd	2.001	1.204
Cr	49.380	29.822
Fe	18.643	31.618
Mn	1,671.035	2,204.680
Hg	1.555	1.251
Ni	29.154	18.438
Pb	274.616	194.900
Cu	37.558	20.903
Zn	683.213	557.930

Table 6: Results of flotation test B (Denver cell, reagent: pine oil+ amyl xanthate)

Products (B)	Float	Sink
Weight	5.4 (%)	94.6 (%)
Metals	ppm	
As	39.680	32.360
Ba	104.370	178.598
Cd	0.878	0.103
Cr	178.069	102.105
Fe	128,049.000	196,137.000
Mn	1.676.700	2.293.692
Hg	2.060	0.280
Ni	32.741	18.769
Pb	403.148	190.347
Cu	44.880	21.517
Zn	789.024	574.542

Fig. **5** and Fig. **6** show spectrophotometric analysis results, respectively for flotation test A and test B.

Spectra obtained from flotation test A and B have a similar graph: *spectra* of sink and float products are placed respectively at the top and in the bottom of the chart; the sediment before treatment is between them, closer to the sink.

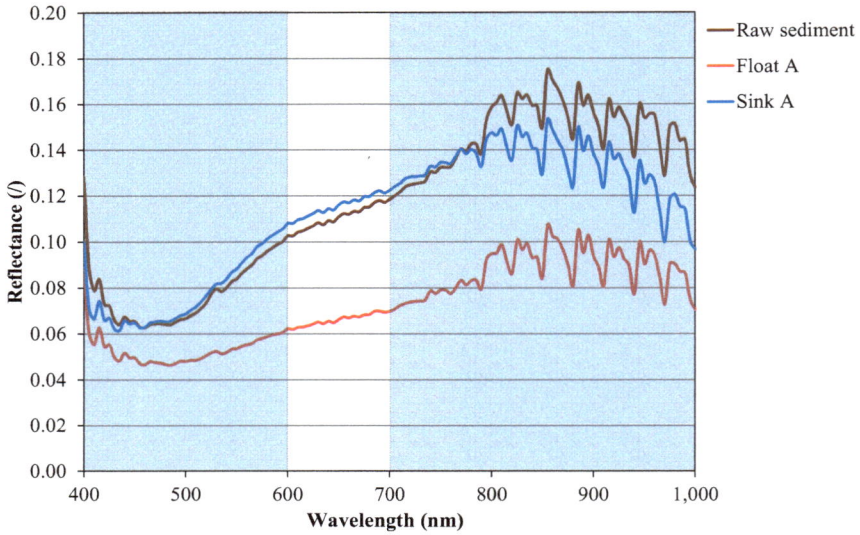

Figure 5: Flotation test A –*Spectra.*

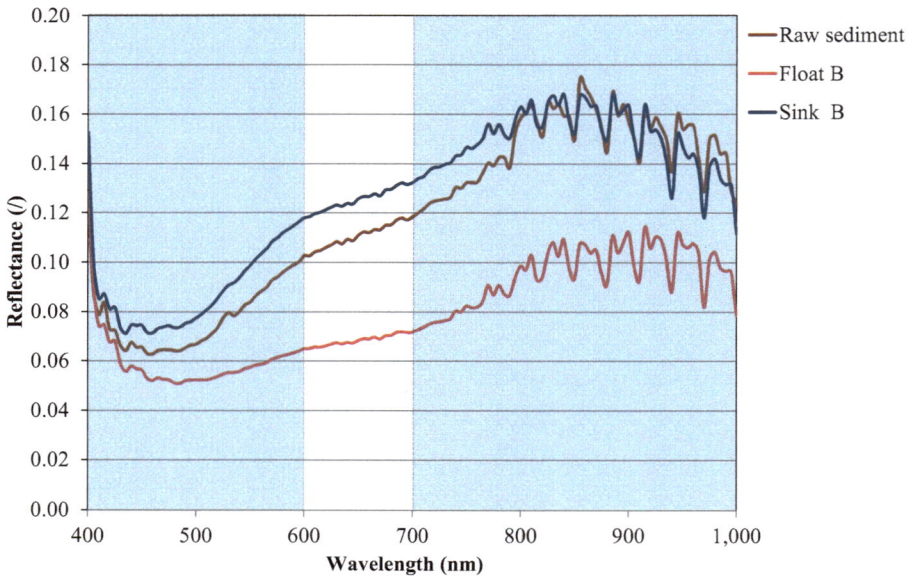

Figure 6: Flotation test B –*Spectra.*

CONCLUSIONS

This experimental research aimed to evaluate the potential of hyperspectral analysis as a fast tool to determine the contamination degree of contaminated sediments.

Spectra relating to the products of each treatment test are compared in Fig. **7**. Notably, products with higher concentration values show similar trends, mostly located in the lower part of the graph. Likewise, products with lower concentration values share a similar behaviour while being mainly located in the higher part of the graph.

The first group of curves has the Overflow *spectra* as lower limit, whereas the second group has the Underflow *spectra* as upper limit. Such feature of Underflow and Overflow *spectra* may be due to the opposite effects of their particle size distribution and metal concentration levels. In fact, Underflow shows high concentration values, similar to Float and -63 μm class, yet its particle size distribution is shifted towards that of Sink and higher particle size classes. The same analysis may be applied to Overflow.

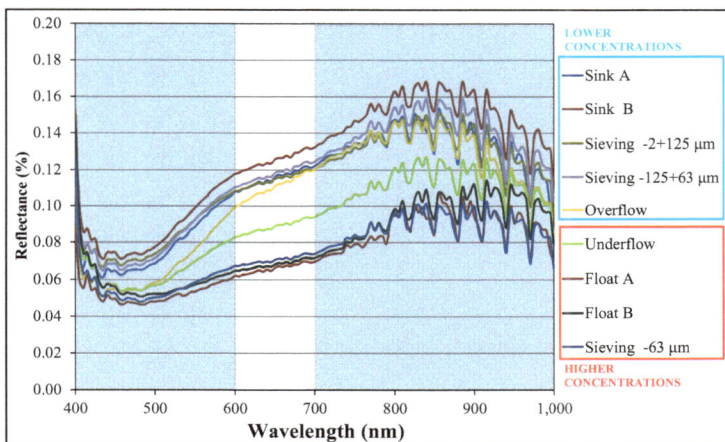

Figure 7: *Spectra* trends.

Inside the medium wavelength range (630-680 nm), a threshold between higher and lower concentration products may be identified for a value of 0.1% of reflectance, as shown in Fig. **8**.

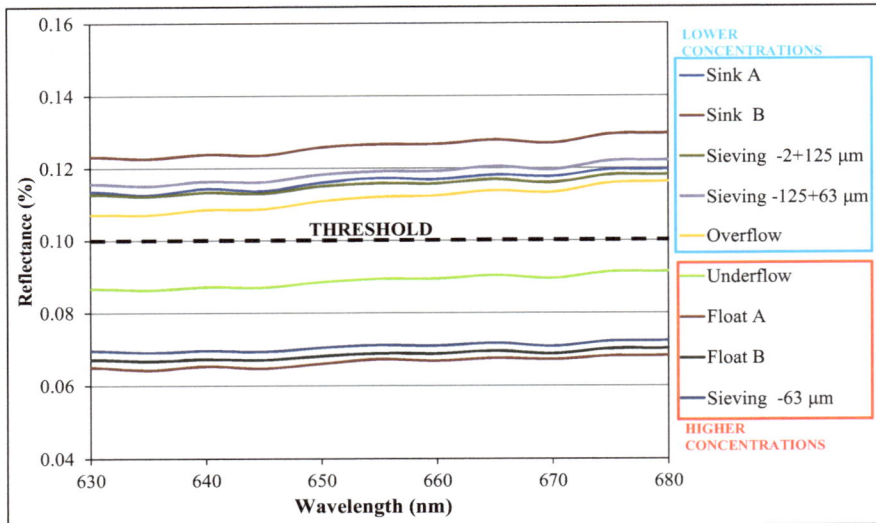

Figure 8: Higher-lower concentration threshold.

Considering the results of the chemical analysis, a correlation can be pointed out between the reflectance curve resulting from hyperspectral analysis and both particle size distribution and contamination degree.

Flotation test *spectra* show similar trends. In fact, as float components contain higher metal concentrations than sink ones, float curves show lower reflectance values than sink's. Moreover, the reflectance curve of raw sediment is placed between sink and float ones. In the same way, sieving test *spectra* show lower reflectance values in the sieve (-63 μm) with the higher metal concentration. Finally, the same reflectance trend is noted for the Hydrocyclone test: as the underflow class contains higher metal concentrations than tailings or overflow, its reflectance curve appears to be lower.

For the feeding sediment samples, results show that *spectra* of raw sediments have higher reflectance values than the -1 mm sieved fraction.

Therefore, the *spectra* trends point out that reflectance values are inversely proportional to the metal contents whilst proportional to the fine fraction content.

In particular, the hyperspectral analysis of the tested sediments samples has pointed out, inside the wavelength range of 630-680 nm, the existence of a reflectance threshold that could be used to assess the residual contamination.

Further studies could be carried out to clarify the connection between *spectra* and characteristics of analysed sediments (*e.g.* hyperspectral analysis could be applied to "artificially contaminated" sediments).

Results of this study highlight the potential of hyperspectral analysis for fast sample characterization during sampling activities, preliminary screening of samples, on line monitoring tool for treatment plants.

ACKNOWLEDGEMENT

Declared none.

CONFLICT OF INTEREST

The author(s) confirm that this chapter content has no conflict of interest.

REFERENCES

Bonifazi, G., Piga, L., Serranti, S., Menesatti, P. (2005). Hyperspectral imaging based techniques in contaminated soils characterization, Optical Sensors and Sensing Systems for Natural Resources and Food Safety and Quality. In: *Proceedings of the SPIE*, 5996, 215-223, Edited by Chen, Yud-Ren; Meyer, George E.; Tu, Shu-I.

Bonifazi, G., La Marca, F., Massacci P. (2002). Characterization of Bulk Particles in Real Time. *Particle and Particle Systems Characterization,19*(4), 240-246.

Cronhjort, B.T., Niemi, A.J., Suoninen, E. (1997). Computer-aided measurements in mining, mineral and metal processing. *Control Engineering Practice*, 5(2),239-245.

Gente, V., Geraldini, S., La Marca, F., Gabellini, M., Palombo, F. (2009). Chemical-Physical Treatments of Marine Contaminated Sediments – A Comparison. *International Journal of Soil, Sediment and Water*, 2(2), 5.

Jones, K.W., Guadagni, A.J., Stern, E.A., Donato, K.R., Clesceri, N.L. (1998). Commercialization of dredged-material decontamination technologies. *Remediation, Special Issue: Innovative Remediation Technology*, 8, 43-54.

Okx, J.P., Stein, A. (2000). An expert support model for *in situ* soil remediation. *Water Air and Soil Pollution,118,*357-375.

Olin-Estes, T.J., Palermo, M.R. (2001). Recovery of dredged material for beneficial use: the future role of physical separation processes. *Journal of Hazardous Materials*, 85, 39–51.

Rivett, M.O., Petts, J., Butler, B., Martin, I. (2002). Remediation of contaminated land and groundwater: experience in England and Wales. *Journal of Environmental Management*, 65(3), 251-268.

Spectral Imaging Ltd (2002). *Imspector User Manual and Application Notes*. Oulu, Finland.

Index

A

Advanced Dry Recovery (ADR) 43, 45, 47-48, 50-51, 54, 57

Acid Mine Drainage (AMD) 140, 141, 143, 148, 150-153, 164, 166

Ammonia 139-142, 144, 148, 151, 156, 158,

B

Biological method/process/treatment 70, 125, 139, 142, 144, 152-154, 156-161, 164

Breakage-matrix 126-128, 137

Bagnoli (Italy) 174

C

Classification 3-5, 8, 12, 19, 21, 24-27, 33, 44, 47, 54, 56-57,

Clean Scrap Machine (CSM) 77, 80-85, 88

Coal power station 3, 22, 24, 28

Concentration criterion 32-34

Construction and Demolition Waste (CDW) 43-45, 47-48, 51, 54, 57, 59, 62, 64, 70, 87-
88, 126,173-174

Contaminated (polluted) sites 27, 123, 173-174

Cyanide 139-142, 156-159, 167

E

End-of-Life Vehicles (ELV) 30, 77-78, 84

F

Flotation 5, 11, 13, 18-20, 30, 32, 34-37, 39-41, 103-105, 141-142, 144, 159, 174-175,
182-184, 186

Fly ash 22-28

G

Granite waste 3, 17-19, 28

Gravity concentration/separation 3, 8-11, 13-14, 21-22, 27, 32-34, 40

Groundwater 44, 123-131, 134-138, 153

H

Hand picking 60, 72, 78, 80, 88

Heavy metals 4-6, 12-13, 74, 144, 153, 156, 163, 178

High density sludge (HDS) process 143, 146

Hyperspectral analysis 172-175, 178, 185-187

I

Incinerator Bottom Ash (IBA) 43-41, 56-57, 77-79, 83-85, 88

In-situ techniques 5

Ion exchange 125, 152, 154-156, 158-159

L

Landfill leachate 139-140, 167

Lime neutralization 161-162

M

Machinery Equipment 63

Membrane process/technology 139, 151, 154-156, 160

Mineral processing 3-7, 20, 27-28, 38, 40, 60, 90, 141

Mine waste 6-9, 11-12

Mining effluents 139-140, 142, 144, 147, 152, 167,

Moist materials 43, 45

Monte Narba mine site (Italy) 6, 27

Montevecchio mine site (Italy) 12-13, 27

N

Near Infrared Spectral Range (separation in) 70,

P

Packaging 30-33, 36-38, 40-41, 60, 70, 74

Particle surface (properties) 32, 34-35, 74

Passive method/process/system/treatment 139, 153-154, 161-165,

Partition function 109, 114-116, 118-119, 121

PET - Polyethylene Terephthalate (separation of) 30, 32-37, 39-41, 69-70

Photovoltaic (PV) module 90-98, 100-103, 105-107

Porto Vesme area (Italy) 20

Post-consumer scrap 78, 88

Precipitation, 4, 97, 105,139, 144-148, 150-153, 156, 158, 160-162, 164, 166

PS - Polystyrene (separation of) 30, 32-34, 37, 39-41, 70

Pump and treat (P&T) 123-125, 129, 136

PVC - Polyvinyl Chloride (separation of) 30, 32-37, 39-41, 70

R

Recycling 27, 30-32, 36, 40, 44, 47, 59-60, 75, 78, 90, 92, 94-95, 97-99, 101-107, 109-111, 114, 117, 120-121, 143, 154, 159, 167-168, 173

S

Sardinian Grey 17

Sediments, 3, 6, 12, 20, 22, 28, 172-175, 178-180, 184-187

Sensor 59-68, 70-75, 77, 79-80, 83-84, 88

Separation efficiency 62, 70, 116

Separation method/process/technique 3-5, 10, 13, 18, 27, 30-36, 40-41, 43-44, 47, 53, 56, 59-63, 66-75, 78, 81, 83-86, 88, 90, 95-98, 100, 102, 104, 109, 111-121, 139-140, 142, 144, 147-148, 151-152, 154-156, 160, 166-167, 174

Sorting 31-32, 59-64, 66, 68-73, 75, 77, 79-80, 84, 88, 112, 119, 174

Spectral range 66, 70, 175

Steel scrap 77-80, 83-84, 88

Sulphate reducing bacteria (SRB) 152-154, 163-165

V

Visible spectral range (separation in) 66, 69

X

X-Ray (sensor) 61, 66-67, 73-74

W

Waste from Electric and Electronic Equipment (WEEE) 30, 77-79, 110

Waste Treatment 3, 59, 70, 75

Waste refrigerators 109-110

Wastewater 138, 138-140, 148, 154-155, 157-158, 165, 167

Wetland 139, 154, 159, 163, 165-166

www.ingramcontent.com/pod-product-compliance
Lightning Source LLC
Chambersburg PA
CBHW041701210326
41598CB00007B/487